UNITED NATIONS ECONOMIC COMMISSION FOR EUROPE

Data for Climate Action: How National Statistical Offices Can Contribute

UNITED NATIONS
Geneva, 2024

ECE/CES/STAT/2024/4

United Nations publication issued by the United Nations Economic Commission for Europe

Sales No: E.25.II.E.3
ISBN: 978-92-1-003374-9
e-ISBN: 978-92-1-106987-7
ISSN: 0069-8458
e-ISSN: 2959-4170

Preface

To address the current and future information demands related to climate action, data producers need to understand the policy context, key stakeholders and their information needs, and be able to communicate effectively how their data can be used for policy applications. This publication aims to provide national statistical offices (NSOs) with such knowledge. It showcases what statistical systems already offer to support climate action, and aims to provide ideas, inspiration and recommendations on how NSOs can make their contribution more valuable for data users.

The publication examines how official statistics can contribute to the reporting under the Paris Agreement; providing data for climate change mitigation, adaptation and just transition; informing the public; and measuring climate-related costs, transfers and financial flows. For each of these topics, it describes the policy context, showcases what NSOs can offer now, identifies gaps they could fill in the future, and provides recommendations on how to better meet user needs. The document does not aim to develop new indicators, classifications, new statistical frameworks and other similar products to avoid duplication of efforts, as a lot of work has already been completed or is currently in progress in various statistical domains. Instead, the document presents how the existing products can be used in the context of climate change and identifies where further development is needed.

While the primary audience of the guidance is NSO staff, both domain experts and managers, the publication may also be useful for data users and other producers of climate change-related statistics and data in countries and international organizations.

The publication was prepared by a dedicated United Nations Economic Commission for Europe Task Force between 2022 and 2024 and endorsed by the Conference of European Statisticians in June 2024.

Acknowledgements

This publication has been produced by members of the Conference of European Statisticians' Task Force on the role of national statistical offices in achieving national climate objectives.

The work was largely undertaken in four groups coordinated by Otto Swertz (Chair of the Task Force, Statistics Netherlands), P. Bhanumati (Group A: Reporting under the Paris Agreement, International Monetary Fund), Arthur Denneman (Group B: Mitigation, Statistics Netherlands and Acting Chair in the first 6 months), Paul McElvaney (Group C: Adaptation, Central Statistics Office of Ireland), Sara Svantesson (Group D: Informing the public, Statistics Denmark).

In addition to the group leads and the Chair, the Task Force included the following members and contributors: Anahit Avetisyan, Nelli Baghdasaryan and Naira Mandalyan (Statistical Committee of the Republic of Armenia), Nona Budoyan, Haykanush Azizyan and Yelena Khalatyan (Ministry of Environment, Armenia), Emil Mammadov (State Statistical Committee of the Republic of Azerbaijan), Oksana Melnikovich (Ministry of Natural Resources and Environmental Protection of the Republic of Belarus), Veronika Zhosan (National Statistical Committee of the Republic of Belarus), Mariam Iqbal, Erin Kumar, Abdoul-Razak Mamane, Joe St. Lawrence (Statistics Canada), Katherine Gómez and Iván Delgado (Ministry of Environment and Energy, Costa Rica), Giovanna Tagliacozzo and Angela Ferruzza (Italian National Institute of Statistics), Assel Shauenova (Bureau of National Statistics of the Agency for Strategic Planning and Reforms of the Republic of Kazakhstan), Muhammad Amjad (Global Climate-Change Impact Studies Centre Pakistan), Beata Nowakowska, Agata Kiełczykowska and Urszula Dawgiałło (Statistics Poland), Anna Olecka (National Centre for Emissions Management (KOBiZE) of Poland), Dušanka Dostanić (Statistical Office of the Republic of Serbia), Antonio Martínez and Ana Solera (National Statistics Institute, Spain), Erhan Ünal (Turkish Statistical Institute), Natalia Husieva (State Statistics Service of Ukraine), Angela Watkins, Tom Sadler and Vijendra Ingole (Office for National Statistics, United Kingdom), Amanda Penistone (Department for Energy Security and Net Zero of the United Kingdom), Christian Schmieder (Bank for International Settlements), Andrew Kanutin (European Central Bank), Ricardo Fernandez and Christiana Photiadou (European Environment Agency), Arturo de la Fuente (Eurostat), Roberta Quadrelli and Pouya Taghavi-Moharamli (IEA), Alessandra Alfieri (IMF), Myriam Linster, Santaro Sakata and Mikaela Rambali (OECD), Karina Miroslava Cazarez, Cathy Krüger, Ana Lucia Londoño, Elisa Narminho and Paz Patino, (PARIS21), Lornaliza Kogler, Maryam Navi and Stuart Best (UNFCCC), Issoufou Seidou (UNECA), Malgorzata Cwiek, Jonathan Gessendorfer and Chaoyi Guo (UNECE), Georgina Alcantar and Alberto Malmierca (UN ECLAC), Rikke Munk-Hansen and Teerapong Praphotjanaporn (UN ESCAP), Reena Shah and Emil Ivanov (UNSD).

Members of the UNECE Steering Group on climate change-related statistics have provided overall guidance to the work of the Task Force and comments on various stages of work. In particular, the authors of this document would like to thank Michele McMillan and Édith Hovington (Statistics Canada), Gerry Brady and Claire O'Hara (Central Statistical Office, Ireland), Stephan Moll (Eurostat), and Tiina Luige and Michael Nagy (UNECE) for comments and contributions to the draft, as well as all experts from countries and international organizations who provided invaluable feedback during the written electronic consultation conducted in April-May 2024, preceding the submission for endorsement to the Conference of European Statistician in June 2024. The Task Force also thanks the participants of 2022 and 2023 UNECE Expert Fora for Producers and Users of Climate Change-Related Statistics for feedback on the draft and examples of their work presented in the Expert Forum or provided through the UNECE Climate Change-Related Statistics in Practice questionnaire, which fed into this document.

The Task Force extends special thanks to journalists who shared their experiences of working with climate-related data, which formed the basis of Chapter 7 of this publication.

Malgorzata Cwiek (UNECE) acted as the Secretary of the Task Force.

Acronyms and abbreviations

ACE	Action for Climate Empowerment
AFOLU	Agriculture, Forestry and Other Land Use
BA	Biennial Assessment and Overview of Climate Finance Flows by the SFC
BTRs	Biennial Transparency Reports under the ETF
CCS	Carbon capture and storage
CES	Conference of European Statisticians
CISAT	Climate Change Statistics and Indicators Self-Assessment Tool of UNSD
CMA	Conference of the Parties serving as the meeting of the Parties to the Paris Agreement
CMFB	Committee on Monetary, Financial and Balance of Payments Statistics of the European Commission
CMP	Conference of the Parties serving as the meeting of the Parties to the Kyoto Protocol
CO_2e	CO_2 equivalent
COP	Conference of the Parties for the Convention
CPA	Classification of Products by Activity
CPC	Central Product Classification
CRT	Common Reporting Tables
CTF	Common Tabular Formats
DGI	Data Gaps Initiative
DRR	Disaster risk reduction
DRSF	ESCAP Disaster-Related Statistics Framework
EEA	European Environment Agency
ESA	European Space Agency
ESG	Environmental, Social and Governance
ETF	Enhanced Transparency Framework established under the Paris Agreement
EU	European Union
EU ETS	European Union Emissions Trading System
EU-SILC	European Union statistics on income and living conditions
EWS	Early warning systems
FAO	Food and Agriculture Organization
FDES	United Nations Framework for the Development of Environment Statistics
GCF	Green Climate Fund
GDP	Gross Domestic Product
GEF	Global Environment Facility
GHG	Greenhouse gas
GIS	Geographic Information System
GST	Global Stocktake
IEA	International Energy Agency
IFCMA	International Forum on Carbon Mitigation Approaches
IMF	International Monetary Fund
IMO	International Maritime Organization
IPAC	International Programme for Action on Climate
IPCC	Intergovernmental Panel on Climate Change
ISIC	International Standard Industrial Classification of All Economic Activities
IUCN	International Union for Conservation of Nature
LULUCF	Land Use, Land-Use Change and Forestry
M&E	Monitoring and evaluation
MPGs	Modalities, Procedures and Guidelines of the Enhanced Transparency Framework
MRV	Measurement, Reporting and Verification
NACE	Statistical Classification of Economic Activities in the European Community
NAP	National Adaptation Plan
NDC	Nationally Determined Contribution
NGO	Non-governmental organizations
NSO	National Statistical Office
NSS	National Statistical System
OECD	Organisation for Economic Co-operation and Development
PM	Particulate matter

RD&D	Research, development and demonstration
REDD+	Reducing emissions from deforestation and forest degradation in developing countries and additional forest-related activities that protect the climate
SBI	Subsidiary Body for Implementation
SBSTA	Subsidiary Body for Scientific and Technological Advice
SCF	Standing Committee on Finance
SDGs	Sustainable Development Goals
SEEA	System of Environmental Economic Accounting
SEEA-CF	System of Environmental Economic Accounting – Central Framework
SEEA-EA	System of Environmental Economic Accounting – Ecosystem Accounting
SNA	System of National Accounts
TF SuFiR	CMFB task force on statistics on sustainable finance and climate related risks
UNDP	United Nations Development Programme
UNDRR	United Nations Office for Disaster Risk Reduction
UNECA	United Nations Economic Commission for Africa
UNECE	United Nations Economic Commission for Europe
UN ECLAC	United Nations Economic Commission for Latin America and the Caribbean
UNEP	United Nations Environment Programme
UNFCCC	United Nations Framework Convention on Climate Change
UN-GGIM	United Nations Committee of Experts on Global Geospatial Information Management
UNSD	United Nations Statistics Division
WHO	World Health Organization
WMO	World Meteorological Organization

Contents

List of tables

List of boxes

List of figures

Chapter 1 Introduction

1. The publication aims to provide guidance on the role of national statistical offices (NSOs) in achieving national climate objectives. It was prepared between February 2022 and January 2024 by a dedicated United Nations Economic Commission for Europe (UNECE) Task Force established by the Bureau of the Conference of European Statisticians and chaired by Statistics Netherlands.

1.1 Background and work of the Task Force

2. Since 2011, the work on climate change-related statistics under the Conference of European Statisticians (CES) has aimed to make official statistics more useful for climate analyses and to promote the involvement of NSOs in greenhouse gas (GHG) inventories and in the climate change domain more broadly.

3. The first UNECE *Recommendations on Climate Change-Related Statistics* (UNECE, 2014), published in 2014, defined the role of NSOs in supporting GHG inventories and producing other climate change-related statistics, and listed many ways how NSOs could increase their involvement in providing climate-related data. A Steering Group on Climate Change-Related Statistics was created to follow up on the implementation of the Recommendations, guide further work in this area and organize annual Expert Fora for Producers and Users of Climate Change-Related Statistics[1] as a platform for collaboration, sharing ideas and experience, discussing concepts and measurement issues, and identifying areas for development of practical guidance. Further, a dedicated Task Force developed a CES set of core climate change-related statistics and indicators (UNECE, 2021a), recommended to be produced in the region to provide a big picture of climate-related issues in an internationally comparable way. In 2016, the work started at the United Nations Statistics Division (UNSD) to develop a Global Set on Climate Change Statistics and Indicators (United Nations, 2024). applicable to countries at various stages of development. The Global Set was adopted at the fifty-third session of the United Nations Statistical Commission in March 2022 as the framework for climate change statistics and indicators to be used by countries when preparing their own sets of climate change statistics and indicators according to their individual concerns, priorities and resources.

4. Many NSOs have been implementing the Recommendations and improving the availability and quality of climate change-related statistics and indicators. However, the complexity of issues related to climate change has been continuously increasing as well. The Paris Agreement introduced new reporting requirements, defined by its 2018 implementation rule book adopted in Katowice. The urgency of action increased, and the policymakers, civil society and researchers started to look more and more into the socio-economic drivers and impacts of climate change. In February 2019, the Bureau of the Conference of European Statisticians recognized that there are significant developments in this domain and decided to conduct an in-depth review of the role of the statistical community in providing data and statistics for climate action (UNECE, 2020). The analysis included an overview of the international activities related to climate change statistics and data, a description of country practices regarding the involvement of NSOs in climate change-related statistics and identified issues and challenges. Among others, the CES Bureau concluded that the topic is strategically important for official statistics, and although NSOs in many countries are actively involved, there is potential for doing more. The official statistics community should increase its visibility in supporting climate action, as it is still often the case that official statistics are not even mentioned in global discussions on climate change-related issues.

5. A subsequent high-level discussion of Chief Statisticians at the 2021 CES plenary session[2] recognized the increasing need for climate change-related statistics and data, and the key role NSOs must play in meeting this need. It was noted that producing such data requires systemic approaches,

[1] https://unece.org/statistics/climate-change/meetings
[2] https://unece.org/info/events/event/348378

linking various statistical fields, innovation, collaboration across the national statistical system, and effective communication with users. The Conference asked for a close collaboration with the policy process, including for developing guidance on using statistics for reporting under the Enhanced Transparency Framework of the Paris Agreement.

6. Responding to that request, the CES Bureau established a Task Force to analyse opportunities for involvement of NSOs in meeting national climate change objectives and showcase what the statistical system already offers to support climate action. The terms of reference of the Task Force are presented in annex 1 to this publication.

7. The Task Force comprised representatives of NSOs of Armenia, Azerbaijan, Belarus, Canada, Costa Rica, Denmark, Ireland, Italy, Kazakhstan, Luxembourg, Netherlands (Chair), Poland, Serbia, Spain, Türkiye, Ukraine and the United Kingdom of Great Britain and Northern Ireland; ministries of environment of Armenia, Belarus and Costa Rica, the National Centre for Emissions Management (KOBiZE) of Poland and the following international organizations: the Bank for International Settlements (BIS), the European Central Bank (ECB), the European Environment Agency (EEA), the International Energy Agency (IEA), the International Monetary Fund (IMF), the Organisation of Economic Co-operation and Development (OECD), the Partnership in Statistics for Development in the 21st Century (PARIS21), the Statistical Office of the European Union (Eurostat), the United Nations Economic Commission for Africa (ECA), the United Nations Economic Commission for Latin America and the Caribbean (ECLAC), the United Nations Economic and Social Commission for Asia and the Pacific (ESCAP), the United Nations Framework Convention for Climate Change (UNFCCC) Secretariat and the United Nations Statistics Division (UNSD).

8. The Task Force worked in four parallel groups tackling the following areas: reporting under the Paris Agreement (Group A), meeting information needs related to climate change adaptation policies (Group B) and climate change mitigation policies (Group C) and informing the broad public (Group D). During the course of its work, the Task Force has identified two additional areas to be included in the publication - data needs related to just transition and climate finance. In addition to the work in parallel groups on thematic areas, the Task Force worked together on developing its work plan and guiding principles, identifying relevant resources and country practices, analysing the institutional landscape, reviewing the draft chapters and identifying cross-cutting issues. The work on the current draft has been carried out through videoconferences and e-mails, and two hybrid meetings in September 2022 and April 2023.

9. In the course of its work, the Task Force has regularly reported to the Steering Group on Climate Change-Related Statistics. In 2022, the Task Force presented to the UNECE Expert Forum for Producers and Users of Climate Change-Related Statistics its work plan, a draft outline of the report and first results from interviews with journalists conducted by Group D. Updates about the work of the Task Force have also been presented to Ninth Meeting of the United Nations Expert Group on Environment Statistics (October 2022), the OECD Working Party on Environmental Information (November 2022) and the Global Webinar on Strengthening Climate Change and Disaster-Related Statistics: Needs, Priorities, and Action organized under the United Nations Development Account project (May 2023). Updates have regularly been shared with the United Nations Committee of Experts on Environmental-Economic Accounting Working Group on Coordination and Communication (Area A).

10. In August 2023, the draft document has been consulted with the participants of the UNECE Expert Forum for Producers and Users of Climate Change-Related Statistics and the Steering Group on Climate Change-Related Statistics. In December 2023, the draft publication was presented at an official COP28 side event in Dubai "Enabling climate action through data, transparency and finance", co-organized by UNECE in Dubai, the United Arab Emirates.

11. The Task Force has built on numerous resources developed under major statistical frameworks and initiatives, such as the Global Set of Climate Change Statistics and Indicators, the Framework for the Development of Environment Statistics, the System of Environmental Economic

Accounting, the G20 Data Gaps Initiative and the OECD International Programme for Action on Climate.

1.2 Target audience

12. This publication has been prepared mainly from the perspective of NSOs. The primary audience of the document are NSO staff, including both domain experts and managers. The document aims to inspire and support people working in NSOs who want to start working in this area or do more.

13. The publication may also be useful for:

(a) Data users by informing about what NSOs could already offer (the primary data users are policymakers, researchers, non-governmental organizations (NGOs) and international organizations);

(b) Other producers of climate change-related data in countries, for example expert institutes, planning authorities, etc.;

(c) International organizations involved in climate change-related statistics.

1.3 Purpose of the document

14. To address the current and future information demands related to national climate objectives, NSOs and other producers of official statistics need to actively engage with primary data users: policymakers, researchers, NGOs and international organizations. To do so, they need to be aware of the policy context, institutional landscape, key stakeholders and their information needs, and be able to communicate effectively how their work is useful for policy applications. This report aims to provide the NSOs with such knowledge, showcase what is already being done, and provide ideas, inspiration and recommendations on how NSOs can further improve their offering and make it more valuable to the data users.

15. For each of the identified areas of involvement, the document:

(a) Summarizes the key, up-to-date available information about policy context and attempts to translate it to the terminology of the primary target audience;

(b) Showcases how NSOs can contribute based on what they can offer now, and identifies gaps which could potentially be filled by NSOs;

(c) Provides a portfolio of real country examples wherever possible;

(d) Provides recommendations how NSOs can better meet user needs.

16. The document does not aim to develop new indicators, classifications, new statistical frameworks and other similar product to avoid duplication of efforts, as a lot of work has already been completed and is currently in progress in various statistical domains. Instead, the document presents how the existing products can be used in the context of climate change and identifies where further development is needed.

1.4 Overview of the document

17. Chapter 2 on the institutional landscape and the role of NSOs provides the big-picture background for the chapters that follow. It describes the global and regional policy frameworks and identifies the most important actors in producing and using climate change-related data. Further, the chapter recalls the traditional strengths of NSOs and the role they play in the national data ecosystems, and how that translates to the climate change domain. Finally, the chapter also introduces the key relevant statistical frameworks and internationally agreed indicator sets.

18. Chapter 3 focuses on the role of NSOs in reporting under the Paris Agreement. From the perspective of the global reporting, data is integral to the implementation of the Paris Agreement. The Agreement relies on pledges (Nationally Determined Contributions), including quantifiable goals tracked through nationally selected indicators (Enhanced Transparency Framework). The collective progress is assessed every five years (the Global Stocktake). By design, countries have a lot of flexibility in setting goals, choosing indicators and setting up reporting arrangements. There are many opportunities for NSOs to potentially contribute, but their role in any of the Paris Agreement processes is not prescribed. The chapter describes the reporting requirements, related data needs, typical steps in the reporting process and the importance of sustainable institutional arrangements. The second part of the chapter shows the entry points for NSOs, explains why their active involvement is important and provides best practices and examples of existing institutional setups in counties.

19. Chapter 4, 5 and 6 give guidance on the role of NSOs in providing data for national policymaking in the areas of climate change mitigation, adaptation and just transition. National governments are putting in place increasingly ambitious climate change mitigation and adaptation policies to deliver on their pledges and contribute to the achievement of the global temperature goal and the global goal on climate change adaptation. High-quality data can help governments design evidence-based, inclusive policies and monitor their success and effects on the economy and society.

20. Chapter 4 on climate change mitigation starts with introducing key definitions and an overview of mitigation policies being put in place in various countries. The second part of the section presents statistics and indicators on emissions, drivers and mitigation efforts that many NSOs already produce, provides ideas of other relevant activities that could be undertaken and identifies the gaps between data supply and demand. Among others, the section explains the importance of good quality energy data and different approaches to estimating GHG emissions.

21. Chapter 5 concentrates on how NSOs can contribute to informing and monitoring climate change adaptation policies. Climate change adaptation is complex and highly context-specific, so the relevant policies may vary greatly depending on the country, region, geographic characteristics of the territory, etc. For this reason, the chapter emphasizes understanding the underlying concepts and definitions, such as risk, vulnerability, hazards and exposure, and outlines the typical national adaptation planning process and its stakeholders to give NSOs the knowledge and tools to navigate their national circumstances. The section also highlights the challenges of assessing adaptation. The second part of the section maps out how NSOs can contribute, both by statistical domain and by type of activity. As the local context is particularly important for climate change adaptation, the section highlights the importance of producing granular, geospatially enabled datasets and supporting local data users.

22. Chapter 6 focuses on data needs related to just transition, that is achieving the transition to low-carbon, resilient world while leaving no one behind. The chapter will unpack the related policy terms and relate them to the NSOs daily work, so that social equity and inclusion in the context of climate change can be mainstreamed into official statistics.

23. Chapter 7 describes the role of NSOs in informing the public about climate-related issues. The Paris Agreement calls for national governments to enhance climate change education, training, public awareness, public participation and public access to information, recognizing the importance of these steps for enhancing global climate action and achieving global climate objectives. Producing timely and relevant climate change-related statistics can be a key contribution of NSOs towards addressing this call and increasing public awareness of climate-related issues. The chapter maps out various audiences and users of climate change-related statistics and their needs, including identifying new and emerging user groups. The chapter also presents an analysis of interviews conducted by the Task Force with journalists from various media outlets and countries, identifying what the public is interested in, what obstacles journalists face when trying to access and use the NSO data and the gaps that have been observed. Further, it describes what NSOs can do to disseminate and communicate climate-related data effectively and contribute to improving public access to information about climate-related issues.

24. Chapter 8 discusses the role of NSOs in measuring climate-related costs, transfers and financial flows which are linked to climate action described in the previous chapters - climate change mitigation, adaptation and just transition. Under the Paris Agreement, there is a separate goal on making financial flows consistent with a pathway towards low GHG emissions and climate-resilient development as well as a requirement for developed countries to provide resources to assist developing countries. At the same time, the financial dimension of climate change is broader. Investors consider companies' stance vis-a-vis climate change and sustainability objectives (energy use, waste management, pollution) - which has led to increase in environmental, social and governance (ESG) reporting. Climate change has also become a key concern for international regulators and a factor for the stability of national and international financial institutions. As the topic is broad and complex, and the official statistical community's contribution is still in fairly early stages, this chapter aims to outline the main issues relevant for statisticians, informs about initiatives already in place and initiates the discussion about the potential role for NSOs in this landscape, complementing the efforts of other actors.

25. Chapter 9 provides guidance on cross-cutting issues identified in previous chapters, such as the need for geospatial data and geospatially enabled data, collaboration with researchers and academia, governance and coordination of climate change-related data, engagement with national policy processes as well as issues around policy impacts on the perceptions, attitudes and behaviours.

26. Chapter 9 provides guidance and examples of good practices on cross-cutting issues identified in previous chapters:

(a) Governance and coordination in producing climate change-related statistics, including establishing sustainable institutional arrangements for the reporting and exploring potential of data stewardship;

(b) Engagement with policymakers through established mechanisms on both statistical side (e.g. statistical advisory council) and policy side (e.g. special bodies, groups or committees dedicated to interdepartmental coordination of climate policy response);

(c) Strengthening data collection, such as examining the traditional data collections, such as censuses and surveys, from the perspective of meeting climate change-related data needs and tapping into new data sources, such as utility meter data;

(d) Need for local and geospatially enabled data, including improving availability of granular/local and geospatially enabled statistical data, and improving interoperability of statistical and geospatial data;

(e) Collaboration with researchers and academia;

(f) Perceptions, attitudes and behaviours as an important enabler of climate action and key component in both mitigation, adaptation and just transition, understanding of which can provide policymakers with key insights for policy design and monitoring.

27. Finally, Chapter 10 summarizes the conclusions emerging from the work of the Task Force, presents recommendations and identifies areas for future work.

28. The document also includes five annexes. Annex 1 includes terms of reference of the Task Force and guiding principles adopted by the Task Force in drafting of the report. Annex 2 is a summary table of coverage of statistical topics by frameworks and approaches. Annex 3 presents an overview of the CES set of core climate change-related indicators. Annex 4 is a list of statistics and indicators included in the Global Set Climate Change Statistics and Indicators. Annex 5 includes individual anonymized replies from journalists interviewed for Chapter 7. Annexes 4 and 5 are only available online.

Chapter 2 Institutional landscape and the role of national statistical offices

30. This chapter presents the main institutional landscape of climate change-related statistics, the role NSOs usually play in it, and provides the foundation for the chapters that follow.

2.1 Institutional landscape

31. Climate change is defined in Article 1 of the United Nations Framework Convention on Climate Change (UNFCCC) as "a change of climate which is attributed directly or indirectly to human activity that alters the composition of the global atmosphere and which is, in addition to natural climate variability, observed over comparable time periods". In addition, the same article also defines the term climate system, which means "the totality of the atmosphere, hydrosphere, biosphere and geosphere and their interactions."

32. The *CES Recommendations on Climate Change-Related Statistics* have defined the scope of climate change-related statistics as: "Environmental, social and economic data that measure the human causes of climate change, the impacts of climate change on human and natural systems, the efforts of humans to avoid the consequences as well as their efforts to adapt to the consequences."

2.1.1 Policy landscape

2.1.1.1 United Nations Framework Convention on Climate Change and the Paris Agreement

33. In 1992, most countries of the world recognized that climate change is a common concern of humankind and adopted the United Nations Framework Convention on Climate Change (UNFCCC, hereafter also referred to as the Convention) aiming to stabilize GHG concentrations at a level that would prevent dangerous anthropogenic (human-induced) interference with the climate system. The Convention entered into force on 21 March 1994. Today, it has near-universal membership. The 197 States and 1 regional economic integration organization[3] that have ratified the Convention are called Parties to the Convention. The Convention divides countries into three main groups according to differing commitments:[4]

(a) Annex I Parties include the industrialized countries that were members of the OECD (Organisation for Economic Co-operation and Development) in 1992, plus countries with economies in transition (the EIT Parties), including the Russian Federation, the Baltic States, and several Central and East European States;

(b) Annex II Parties consist of the OECD members of Annex I, but not the EIT Parties. They are required to provide financial resources to enable developing countries to undertake emissions reduction activities under the Convention and to help them adapt to adverse effects of climate change. In addition, they have to "take all practicable steps" to promote the development and transfer of environmentally friendly technologies to EIT Parties and developing countries. Funding provided by Annex II Parties is channelled mostly through the Convention's financial mechanism;

(c) Non-Annex I Parties are mostly developing countries. Certain groups of developing countries are recognized by the Convention as being especially vulnerable to the adverse impacts of climate change, including countries with low-lying coastal areas and those prone to desertification and drought. Others (such as countries that rely heavily on income from fossil fuel production and commerce) feel more vulnerable to the potential economic impacts of climate change response measures. The Convention emphasizes activities that promise to

[3] https://unfccc.int/process-and-meetings/the-convention/status-of-ratification-of-the-convention
[4] https://unfccc.int/process-and-meetings/what-are-parties-non-party-stakeholders

answer the special needs and concerns of these vulnerable countries, such as investment, insurance and technology transfer;

(d) The 46 Parties classified as least developed countries (LDCs) by the United Nations are given special consideration under the Convention on account of their limited capacity to respond to climate change and adapt to its adverse effects. Parties are urged to take full account of the special situation of LDCs when considering funding and technology-transfer activities.

34. The Convention required the industrialized countries (Annex I Parties) to regularly report their GHG emissions and adopt policies on limiting them but did not have binding limits on emissions for individual countries. The Convention entered into force in 1994.

35. In 1997, the Kyoto Protocol was adopted as a separate agreement under the Convention (UNFCCC, 1997). The Kyoto Protocol operationalizes the United Nations Framework Convention on Climate Change by committing the Annex I Parties to limit and reduce GHG emissions in accordance with agreed individual targets. The Kyoto Protocol entered into force in 2005. The Doha Amendment to the Kyoto Protocol was adopted in 2012 and entered into force on 31 December 2020.

36. At the 21st Conference of the Parties to the Convention (COP21) in Paris in 2015, 195 countries adopted an agreement on accelerating and intensifying the efforts to combat climate change - the Paris Agreement (UNFCCC, 2015)[5]. The Paris Agreement aims to:

> *"Strengthen the global response to the threat of climate change, in the context of sustainable development and efforts to eradicate poverty, including by:*
>
> *(a) Holding the increase in the global average temperature to well below 2°C above pre-industrial levels and pursuing efforts to limit the temperature increase to 1.5°C above pre-industrial levels, recognizing that this would significantly reduce the risks and impacts of climate change;*
>
> *(b) Increasing the ability to adapt to the adverse impacts of climate change and foster climate resilience and low greenhouse gas emissions development, in a manner that does not threaten food production; and*
>
> *(c) Making finance flows consistent with a pathway towards low greenhouse gas emissions and climate-resilient development."*

Figure 2.1.
Goals of the Paris Agreement

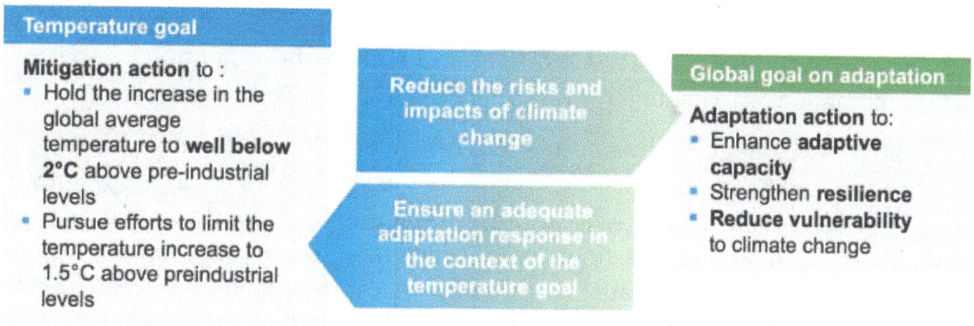

***Source*: UNECE (2020) adapted from UNEP (2017b).**

37. The Paris Agreement requires all Parties to put forward their best efforts to address climate change through "nationally determined contributions" (NDCs) and to strengthen these efforts in the years ahead. This includes the requirement that all Parties regularly report on their emissions and implementation efforts. In their NDCs, countries can include domestic adaptation goals, which together with national adaptation plans (NAPs) allow the countries to design and adjust their adaptation trajectories as their needs evolve and as progress is evaluated over time.

[5] See the dedicated UNFCCC web page https://unfccc.int/process-and-meetings/the-paris-agreement

38. The Convention, the Kyoto Protocol and the Paris Agreement define the institutional arrangements for the climate change intergovernmental process. The arrangements include the following bodies:

(a) A supreme governing body: the Conference of the Parties (COP) for the Convention, the Conference of the Parties serving as the meeting of the Parties to the Kyoto Protocol (CMP), and Conference of the Parties serving as the meeting of the Parties to the Paris Agreement (CMA);

(b) A process management body: the Bureau of COP, CMP and CMA;

(c) Subsidiary bodies:

  (i) Subsidiary Body for Scientific and Technological Advice (SBSTA);

  (ii) Subsidiary Body for Implementation (SBI);

  (iii) Other ad hoc subsidiary bodies established by COP, CMP, or CMA;

(d) Technical subsidiary bodies with limited membership (referred to as the constituted bodies), such as the Adaptation Committee, Adaptation Fund Board, Advisory Board of the Climate Technology Centre and Network, Executive Board of the Clean Development Mechanism, Compliance Committee, Consultative Group of Experts, Executive Committee of the Warsaw International Mechanism for Loss and Damage.

(e) The UNFCCC Secretariat (UN Climate Change);

(f) Entities entrusted with the operations of the financial mechanism: the Global Environment Facility (GEF) and the Green Climate Fund (GCF).

2.1.1.2 2030 Agenda for Sustainable Development

39. The 2030 Agenda for Sustainable Development (United Nations, 2015) called for "taking urgent action to combat climate change and its impacts", and "promoting mechanisms for raising capacities for effective climate change-related planning and management", which is included in goal 13, encompassing the following targets:

(a) 13.1: Strengthen resilience and adaptive capacity to climate-related hazards and natural disasters in all countries;

(b) 13.2: Integrate climate change measures into national policies, strategies and planning;

(c) 13.3: Improve education, awareness-raising and human and institutional capacity on climate change mitigation, adaptation, impact reduction and early warning;

(d) 13.A: Implement the commitment undertaken by developed-country parties to the United Nations Framework Convention on Climate Change to a goal of mobilizing jointly $100 billion annually by 2020 from all sources to address the needs of developing countries in the context of meaningful mitigation actions and transparency on implementation, and fully operationalize the Green Climate Fund through its capitalization as soon as possible;

(e) 13.B: Promote mechanisms for raising capacity for effective climate change-related planning and management in least developed countries and small island developing States, including focusing on women, youth and local and marginalized communities.

40. Combatting climate change and its impacts is one of the sustainable development goals - Goal 13 - but it is intrinsically linked to all other 16 Goals of the 2030 Agenda for Sustainable Development. If humanity fails to halt climate change and put in place measures reducing its impacts, the remaining goals cannot be achieved. SDG 13 is specifically linked to Goal 12: "Ensure sustainable consumption and production patterns"; Goal 9: "Build resilient infrastructure, promote inclusive and sustainable industrialization and foster innovation" and Goal 7: "Ensure access to affordable, reliable, sustainable and modern energy for all". The 2019 SDG progress report pointed to various areas across

the Sustainable Development Goals (SDGs) where lack of progress on climate action is undermining hard fought gains in other areas, including food security (SDG 2) (United Nations, 2019).

41. The Global Indicator Framework for measuring progress on the implementation of Agenda 2030 is described in section 2.3.10.

2.1.1.3 Sendai Framework for Disaster Risk Reduction 2015-2030

42. The Sendai Framework for Disaster Risk Reduction 2015-2030 (hereafter the Sendai Framework) was adopted at the Third United Nations World Conference in Sendai, Japan, in March 2015 (UNISDR, 2015). It is the outcome of stakeholder consultations initiated in March 2012 and inter-governmental negotiations from July 2014 to March 2015, supported by the United Nations Office for Disaster Risk Reduction (UNDRR) at the request of the UN General Assembly. The Sendai Framework has seven global targets. These targets are measured at the global level with 38 indicators. National targets and indicators will contribute to the achievement of the outcome and goal of the present Framework. The seven global targets are:

(a) Substantially reduce global disaster mortality by 2030, aiming to lower the average per 100,000 global mortality rate in the decade 2020-2030 compared to the period 2005-2015;

(b) Substantially reduce the number of affected people globally by 2030, aiming to lower the average global figure per 100,000 in the decade 2020-2030 compared to the period 2005-2015;

(c) Reduce direct disaster economic loss in relation to global gross domestic product (GDP) by 2030;

(d) Substantially reduce disaster damage to critical infrastructure and disruption of basic services, among them health and educational facilities, including through developing their resilience by 2030;

(e) Substantially increase the number of countries with national and local disaster risk reduction strategies by 2020;

(f) Substantially enhance international cooperation to developing countries through adequate and sustainable support to complement their national actions for implementation of the present Framework by 2030;

(g) Substantially increase the availability of and access to multi-hazard early warning systems and disaster risk information and assessments to people by 2030.

Figure 2.2.
Seven global targets of the Sendai Framework

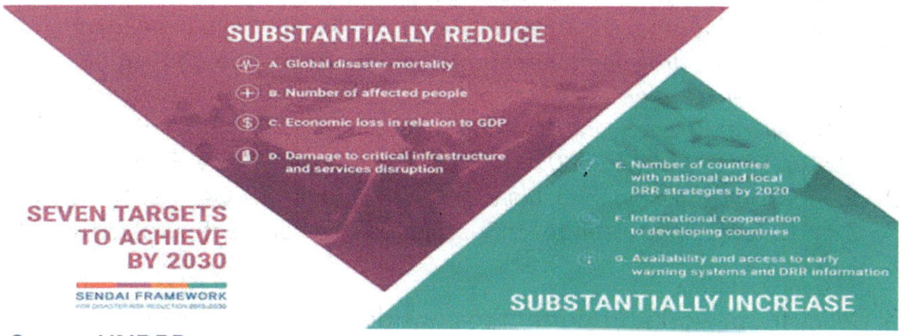

Source: UNDRR.

43. On 3 June 2015, the United Nations General Assembly adopted the establishment of the Open-ended Intergovernmental Expert Working Group on indicators and terminology relating to disaster risk reduction (OIEWG) via resolution 69/284.[6] In three meetings held between September 2015 and November 2016, OIEWG developed the set of indicators to measure global progress in the

[6] https://www.preventionweb.net/terminology/open-ended-working-group

implementation of the Sendai Framework and agreed on the related terminology. More information on the Sendai Framework indicators is presented in section 2.3.9.

Figure 2.3.
Links between the 2030 Agenda, the Paris Agreement and the Sendai Framework for Disaster Risk Reduction

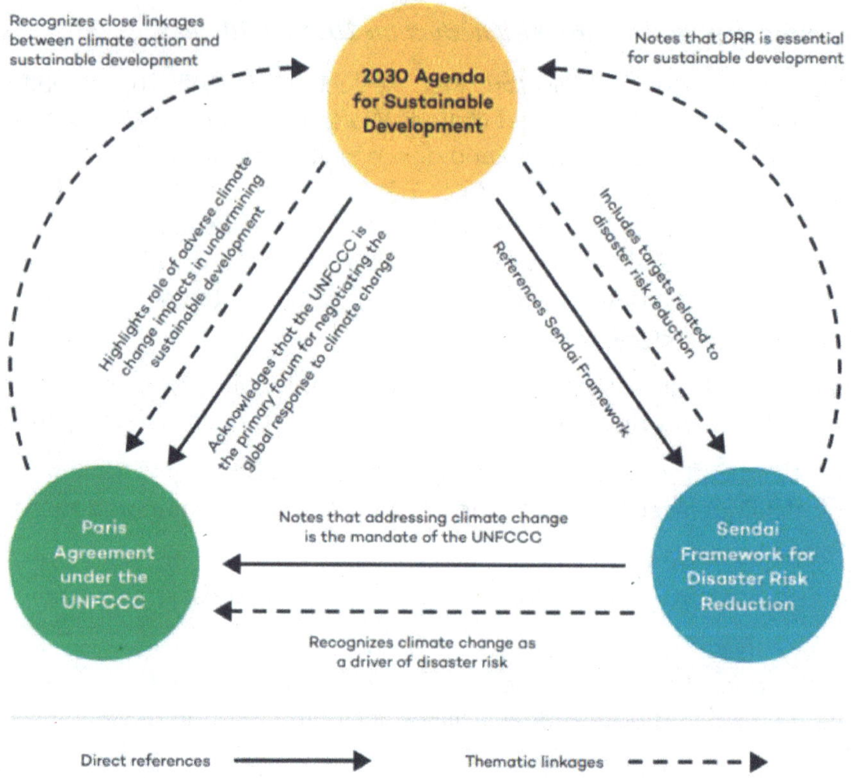

Source: GIZ and IISD (2018).

Box 2.1. Links between the 2030 Agenda, the Paris Agreement, and the Sendai Framework

"Sustainable development, disaster risk reduction and climate change adaptation are inextricably linked and interconnected fields (see Figure 2.3). Building strategic linkages and coordinating actions among them could increase the effectiveness of different actions and related goals. In 2015, the global community adopted the 2030 Agenda for Sustainable Development, the Paris Agreement and the Sendai Framework on Disaster Risk Reduction (DRR). All of them have substantial areas of overlap and clear convergence of objectives related to strengthening resilience, fostering sustainable development and reducing vulnerability to climate change and disasters. For example, the Paris Agreement recognizes the links between climate action and sustainable development while the 2030 Agenda highlights the role of climate change impacts in undermining sustainable development. The Paris Agreement has significant overlap with the Sendai Framework when it deals with losses and damage caused by climate change and extreme weather. Furthermore, Sendai recognizes climate change as a driver of disaster risk and that DRR is essential for sustainable development. In addition, all three frameworks have formulated improved cooperation and international support as a common goal (in SDG 17 of the 2030 Agenda and in both the Sendai Framework and Paris Agreement), creating the foundation for creating functional linkages." (Deutsche Gesellschaft für Internationale Zusammenarbeit (GIZ) and International Institute for Sustainable Development (IISD), 2018)

44. In addition to the global policy frameworks, important to consider are also the regional policy initiatives. In the UNECE region, the most prominent example relevant for many UNECE member States is the European Green Deal (European Commission, 2019).

45. In 2019, a new European Commission took office under the presidency of Ursula von der Leyen and adopted as the first of its priorities a European Green Deal. This is a roadmap for EU legislative and non-legislative initiatives by 2024 to transform the European Union into a modern,

resource-efficient and competitive economy. One flagship initiative was the European Climate Law, setting targets for climate neutrality by 2050 and intermediate targets for 2030.

Figure 2.4.
European Green Deal

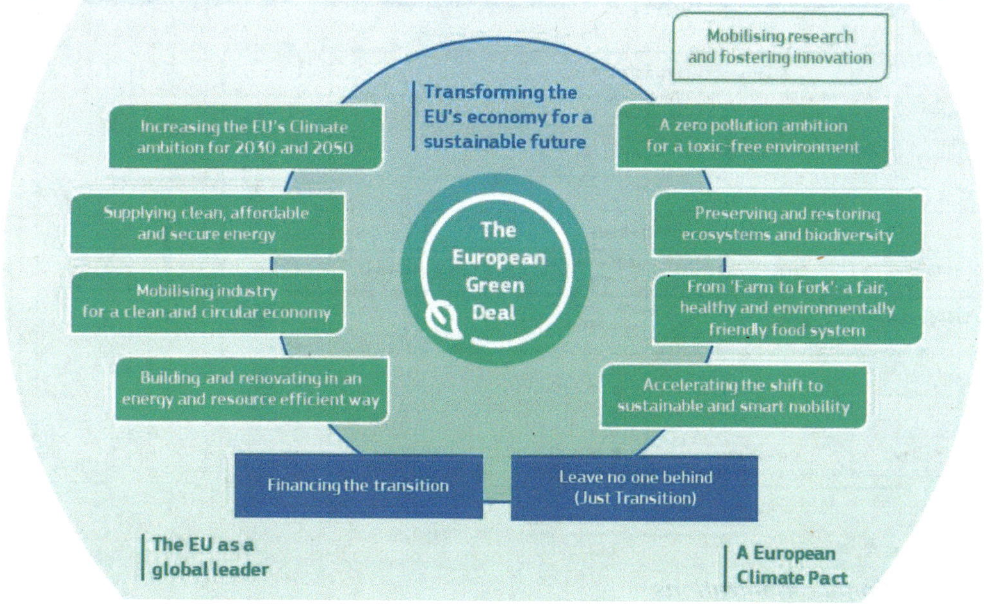

Source: European Commission (2019).

46. The European Statistical System (ESS), under the leadership of Eurostat, adopted in 2021 an Action plan of statistics for the European Green Deal (ESSC, 2021). The Action plan coordinates and streamlines current statistical activities related to the Green Deal. It consists of 29 activities organized in the following strands of work: enhanced communication, update the legal basis for European official statistics, and develop statistics beyond the legal basis (e.g. experimental statistics). About 2/3 of the activities concern environment, climate change, energy, transport and geo-localized statistics; the remaining ones are about agricultural, business or social statistics. All the activities are planned to be completed by end 2024.

2.1.2 Actors in the institutional landscape related to climate change

47. The institutional landscape related to climate change includes actors from diverse stakeholder groups at international, national and local levels (see Figure 2.5).

Figure 2.5.
Examples of actors in the institutional landscape clustered by main stakeholder group

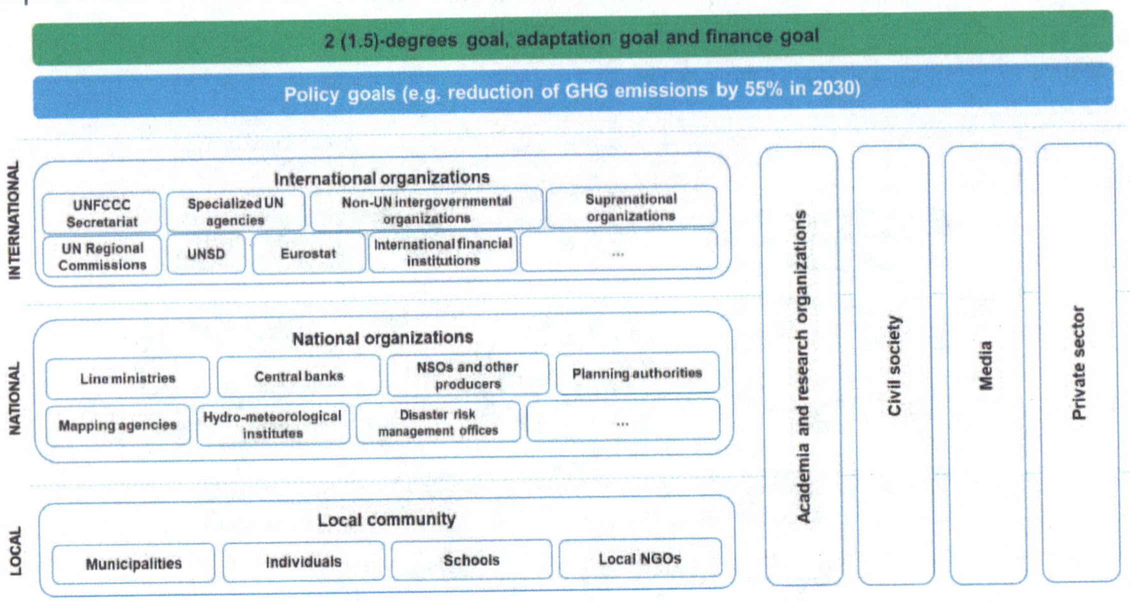

Source: Authors.

2.1.2.1 International organizations

48. Climate change is affecting all areas of human activity, climate change considerations are increasingly mainstreamed into all areas of development work.[7] As identified in the CES in-depth review of the role of statistical community in climate action, the high number of international organizations, researchers, non-governmental organizations working on issues of climate change across all thematic areas reflects the recognition of the importance and multidimensionality of this challenge.

49. The climate-related work of international organizations often includes activities related to data, statistics and measurement. Many international organizations are involved in the production and use of climate change-related information, with their expert networks and national counterparts. The work of international organizations (and their national counterparts) could be grouped as follows (groups are not mutually exclusive):

(a) Custodian of the global policy process: UNFCCC collects official data and information that is relevant for and used to inform various global political processes;

(b) Production and dissemination of data on climate and earth physical systems: e.g. WMO, NASA, Copernicus;

(c) Production and dissemination of climate change-related environmental data, statistics and indicators: e.g. UNEP, IUCN, Eurostat, EEA and the Food and Agricultural Organization of the United Nations (FAO).

(d) Production of analysis and knowledge products that contribute to the global political processes and support national governments in how they can deliver on their commitments to climate action, e.g. World Bank, UNDP, OECD, FAO, IEA, International Renewable Energy Agency, UN regional commissions.

(e) Specialized UN agencies and other international organizations further the understanding of climate change drivers, impacts and adaptation options in their specific domains and contribute to developing national capacities in this regard, such as IAE, FAO, UNDRR or WHO.

---

[7] CES In-depth review on the role of the statistical community in climate action (2020)

(f) UNSD, the statistical divisions of UN regional commissions with their respective expert groups and teams of specialists, and others (e.g., UNEP or PARIS21) support the national statistical systems through development of guidelines, recommendations and other resource,s and supporting their implementation to improve the availability of relevant, harmonized data and increase their use for decision-making.

2.1.2.2 National institutions

50. For achieving national climate objectives, key stakeholders are at the national level and include many producers of data and statistics related to climate change:

(a) Line ministries including national focal points, energy ministries and energy agencies, etc.;

(b) Environmental agencies and environment protection institutes;

(c) NSOs;

(d) Other producers of official statistics, (members of national statistical systems), for example, statistical entities in the ministries of environment, agriculture, energy and others;

(e) Hydrometeorological institutes;

(f) Disaster risk management authorities.

51. **Development cooperation agencies** are also relevant actors as they may be involved in many activities related to climate action in other countries – providing funding, supporting implementation, and monitoring and evaluation of the activities they support. The relevance and scale of these activities is increasing with mainstreaming of climate considerations into development activities and increasing mobilization of climate finance.

52. Existing mechanisms for coordination among the producers should be used to ensure the coherence of activities and outputs. The production of statistics and indicators requires cooperation among various domain experts, especially in the areas of impacts and adaptation, where the domain expertise needs to be combined with an understanding of how climate change is affecting a given area.

2.1.2.3 Local actors

53. Local actors are both affected by climate change and recognized as key agents of climate action. These may include local governments/institutions and cities; communities; civil society, individuals or schools.

2.1.2.4 Academia and the scientific community

54. The **Intergovernmental Panel on Climate Change** (IPCC)[8] is the United Nations body for assessing the science related to climate change and the key global interface between climate science and policy. It has been established by UNEP and WMO at the request of the United Nations General Assembly. IPCC does not conduct its own research, but it assesses the scientific basis of climate change, its impacts and future risks, and options for adaptation and mitigation based on published literature. IPCC identifies where there is agreement in the scientific community, where there are differences of opinions and where further research is required.

55. IPCC assessments provide a scientific basis for governments at all levels to develop climate related policies and underlie the negotiations at the UN climate conferences. **IPCC works in assessment cycles and produces** three types of reports:

(a) Assessment reports, e.g. the Sixth Assessment Report (IPCC, 2022a, 2022b and 2023a);

(b) Special reports, e.g. *Global Warming of 1.5°C* (IPCC, 2018);

---

[8] https://www.ipcc.ch/

(c) Methodology reports, e.g. *2019 Refinement to the 2006 IPCC Guidelines for National Greenhouse Gas Inventories* (IPCC, 2019).

56. Assessment reports cover the full assessment of the state of scientific, technical and socio-economic knowledge on climate change, its impacts and future risks, and options for reducing the rate at which climate change is taking place. Special reports are assessments of a specific issue agreed by its member governments. Methodology reports provide practical guidelines for the preparation of GHG inventories under the UNFCCC (see more details in section 2.3.4). Since 1988, IPCC has produced six comprehensive Assessment Reports and several Special Reports on specific topics.

57. **Universities, scientific institutes, research organizations and collaborations** may among others play the following roles, relevant in the context of this report:

(a) Address crucial scientific questions in the fields of global change, climate impacts and sustainable development;

(b) Bring together researchers from various branches of science to generate interdisciplinary insights and to provide policymakers and society with sound information for decision making;

(c) Share existing national research capacities and infrastructures;

(d) Track progress on certain areas, and provide an independent assessment of the delivery of commitments made by governments worldwide under the Paris Agreement;

(e) Linking science and policy analysis to address human-induced climate change, research and implementation of action against climate change around the globe;

(f) Produce climate-change related data that complements existing official statistics or to close specific data gaps.

58. **Think-tanks** work with governments and non-governmental organizations to support governments, communities and industries to advise on solutions and address key challenges. They may also undertake research and analysis to generate evidence and promote evidenced-informed policies.

2.1.2.5 Non-governmental organizations and other

59. There are also many **non-profit, non-governmental organizations, civil society organizations**, often with international reach, which aim to promote transparency in the climate action and hence are important data users. In some cases they can also act as data producers:

(a) Bring visibility to how climate change affects local communities and vulnerable groups;

(b) Advocate for climate change initiatives;

(c) Produce citizen-generated data that can be used to close specific gaps for reporting on climate change.

2.1.2.6 Private sector

60. Private sector plays several important roles in the context of climate change. For example, choices made by private sector actors in their everyday business contribute to the collective progress (or lack thereof) in addressing climate crisis and, broader, sustainable development. Private sector provides technologies, products and services necessary for mitigation or adaptation solutions and is a key source of innovations that can unlock climate action in new areas. Moreover, private sector actors themselves face climate risk and the risks associated with the transition to low-carbon economy that they need to address. Finally, the private sector can also be a key source of financing of climate action and producer of climate change-related data.

2.1.2.7 Media

61. **Media and journalists** use climate change-related data and statistics to inform the public about a broad range of climate change-related issues and create awareness around climate change and its impacts on society. The topics may cover a broad set of interconnected issues spanning environmental, social and economic topics, covering the wider environment, energy, agriculture, transport, business and politics. The focus can vary from small-scale local action to global long-term trends and comparisons. (See also 7.2.2.4 Experiences on the interest and information needs of the public).

2.1.3 Relationship between policy, science and data

62. In the climate change domain there is a lot of emphasis on the relationship between policy and science. From the perspective of the purpose of this report, it is also useful to bring into the picture the third category - statistics and data, which is the foundation of both sound decision-making and scientific research.

63. Figure 2.6 below illustrates the bidirectional relationships between policy, science and statistics and data in the climate change domain.

Figure 2.6.
Relationship between policy, science, and statistics and data in the climate change domain

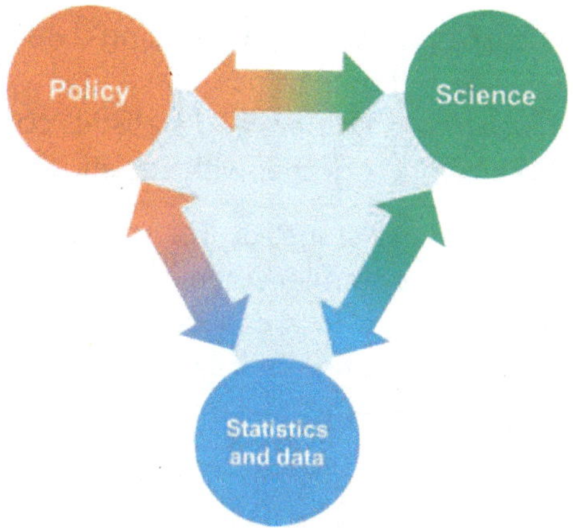

Source: Authors.

64. **Science** provides the basis for defining and formulating policy goals, choosing course of action, and checking their effectiveness. **Policy** can formulate questions to be answered by **science**. This includes model-based scenarios to check if measures are effective.

65. **Policy** defines information needs and priorities for **statistics**. **Data and statistics** provide basis for evidence-based action, monitoring of its effects and reporting on the goals of **policy**.

66. **Science** provides methods and ideas for producing new **statistics**. **Data and statistics** are used by **science** for research, both as aggregates and microdata.

2.1.4 Roles in producing climate change-related statistics data

67. The roles in producing climate change related data range from data holders to statistics producers and data users (Figure 2.7). The different actors described above can take on each of the roles in the context of the same or different type of data. For example, NSOs can take on the role as data holder and statistics producer, and an energy agency can take the role as data holder as well as data user. Businesses can be a data subject, data holders and data users. Individuals are data subjects and data users, etc.

Figure 2.7
Roles in producing climate change-related statistics data

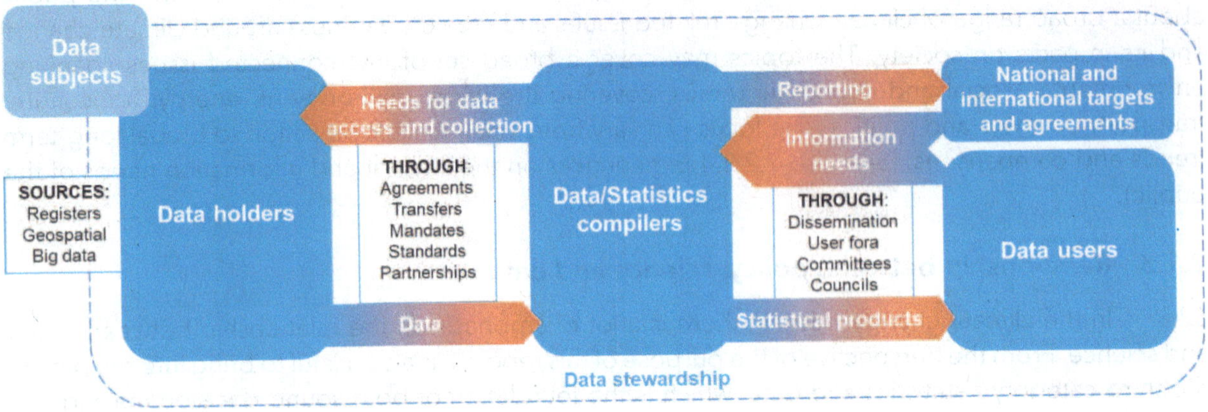

Source: Authors.

68. The role of **data holder or data provider** covers collecting, receiving or having access to different types of climate change related data. Data holders will receive the source data of different types, including survey data, register or administrative data, geospatially enabled data as well as big data.

69. **Data/Statistics compilers** are responsible for compiling data and statistics including reporting official climate change related statistics and data to national and international targets or agreements. Mechanisms and agreements between data holders and statistics compilers will ensure efficient data transfers. Potentially, data holders can work together with statistics producers in expert groups.

70. **Data users** are all users of data and statistics. This can be the public, experts or other user groups (see Chapter 7). The statistics compiler communicates with the data users regarding dissemination of statistics and analyses. User fora can be a way of the two roles working together. Beyond these roles, actors listed in the institutional landscape can also contribute to mobilizing resources and developing capacity for better production and use of climate change-related data.

2.2 Role of the NSO

2.2.1 General characteristics of NSOs and national statistical systems[9]

71. The NSOs in most countries are the main part of the national statistical system (NSS) comprising the organizations and units in a country that collect, process and disseminate official statistics on behalf of the government.

72. In general, official statistics are statistical outputs produced by an NSO or other members of NSS. Usually, official statistics describe the economy, demography, environment, social and cultural situations in a particular country, and are produced within the scope of the statistical programme which guides the work of the national statistical system.

Fundamental Principles of Official Statistics

73. The term "official statistics" serves as a quality label, indicating that official statistics are developed, produced and disseminated in accordance with internationally agreed statistical standards and recommendations, the Fundamental Principles of Official Statistics (see Box 2.2) and the European Statistics Code of Practice (in the EU) or OECD Recommendations on Good Statistical Practice.

[9] The content of this section is based on existing UNECE materials concerning role of NSOs and strengths of official statistics. For more details, please see UNECE (2022), UNECE (2019b) and UNECE (2023a).

Box 2.2.
United Nations Fundamental Principles of Official Statistics

1. Relevance, impartiality and equal access: Official statistics provide an indispensable element in the information system of a democratic society, serving the government, the economy and the public with data about the economic, demographic, social and environmental situation. To this end, official statistics that meet the test of practical utility are to be compiled and made available on an impartial basis by official statistical agencies to honour citizens' entitlement to public information.

2. Professional standards and ethics: To retain trust in official statistics, the statistical agencies need to decide according to strictly professional considerations, including scientific principles and professional ethics, on the methods and procedures for the collection, processing, storage and presentation of statistical data.

3. Accountability and transparency: To facilitate a correct interpretation of the data, the statistical agencies are to present information according to scientific standards on the sources, methods and procedures of the statistics.

4. Prevention of misuse: The statistical agencies are entitled to comment on erroneous interpretation and misuse of statistics.

5. Sources of official statistics: Data for statistical purposes may be drawn from all types of sources, be they statistical surveys or administrative records. Statistical agencies are to choose the source with regard to quality, timeliness, costs and the burden on respondents.

6. Confidentiality: Individual data collected by statistical agencies for statistical compilation, whether they refer to natural or legal persons, are to be strictly confidential and used exclusively for statistical purposes.

7. Legislation: The laws, regulations and measures under which the statistical systems operate are to be made public.

8. National coordination: Coordination among statistical agencies within countries is essential to achieve consistency and efficiency in the statistical system.

9. Use of international standards: The use by statistical agencies in each country of international concepts, classifications and methods promotes the consistency and efficiency of statistical systems at all official levels.

10. International cooperation: Bilateral and multilateral cooperation in statistics contributes to the improvement of systems of official statistics in all countries.

Relevance, impartiality and professional independence

74. Official statistics are an indispensable element in the information system of democratic societies, serving governments, economies and the public with data about the economic, demographic, social and environmental situation, which can help them make decisions.

75. Utility and relevance of official statistics are ensured through established mechanisms for user representation, for instance a Statistical Advisory Council, or other consultations carried out related to the planning of statistical work or to gather feedback on user satisfaction.

76. The credibility of NSOs builds on the trust of users in official statistics as an objective source of information that does not serve the interest of any specific group. Public trust in statistics relies heavily on the strict adherence by statistical offices to scientific, confidentiality and impartiality principles and the use of internationally agreed methodologies. A cornerstone principle of official statistics is impartiality and professional independence, meaning that producers of official statistics shall decide, independently and free from any pressures or interference from political or other external sources, based on sound statistical methodology only, on the development, production, dissemination and communication of statistics. This includes the selection of data sources, concepts, definitions, methods and classifications to be used, and the timing and content of all forms of dissemination and communication. Impartiality of official statistics contributes to a wider social goal of fostering transparency and accountability, building an open relationship between society and states.

Institutional setting, legislation, mandate for data collection and confidentiality

77. NSO usually reports either to a ministry or to the parliament, is financed mainly by the state budget and operates according to the national legal framework. In most countries, there is a separate law on official statistics, which covers the main principles and definitions of official statistics, organization of the NSS, any advisory bodies, coordination of the NSS, statistical programmes, mandate for data collection and access to data, statistical confidentiality, quality of official statistics, dissemination and communication of statistics, provision of statistical services, international cooperation, infringements and relationship to other legislation.

78. The statistical law gives NSOs a clear mandate to collect data from respondents, which allows for the provision of comprehensive, accurate and high-quality statistics. In addition, to function efficiently, the producers of official statistics should have the authority to access administrative data sources existing in the country, free of charge, and to exchange data and metadata within the national statistical system for statistical purposes. Increasing data needs in society, not least due to the reporting on the 2030 Agenda, call for an extended mandate to collect data, including from private data holders.

79. To balance the strong mandate for data collection, the law puts on the NSO a responsibility to inform respondents, monitor and limit response burden and, crucially, protect the confidentiality of the collected data. The individual information is both legally and technically protected and there are strict rules how such information can be used and potentially shared.

Coordination and statistical programmes

80. As mentioned above, beside the NSO, other public agencies can be designated as producers of official statistics, such as central banks and government ministries. The NSO is, in many countries, tasked to coordinate this system of producers of official statistics - the national statistical system, with the aim of meeting the relevant quality standards and the main principles of official statistics. Effective coordination is key to the cost-efficient and consistent functioning of an NSS that demonstrates high professional standards.

81. Statistical programmes are the main instruments for such coordination in most countries. The purpose of statistical programmes is to plan activities of the NSS and efficiently coordinate work between the producers of official statistics. In some countries, the statistical programme provides the mandate for receiving the budget, carrying out statistical surveys, accessing specified data sets and compiling certain statistics. In this case, the programme legitimates data collection from respondents and formalizes administrative data provision. Statistical programmes can also be a tool for designating producers of official statistics by involving them in the preparation and listing them in the work programme. In some countries, the statistical programmes may cover only activities performed by the NSO.

Did you know?

The coordination role of NSOs in the context of statistics for the Sustainable Development Goals has been formalized in the UN General Assembly Resolution on the Work of the Statistical Commission pertaining to *the 2030 Agenda for Sustainable Development* (A/RES/71/313). The resolution "stresses the role of national statistical offices as the coordinator of the national statistical system".

Financing and allocation of resources

82. The NSO budget should cover the overall production of statistics and staff costs, but also the necessary funds for investing in infrastructure, new technologies and new statistical methodologies, etc. However, statistical offices are public institutions financed by state budgets and may have limited mandate in resource management.

83. Typically, financial and human resources are managed in accordance with other legislation that regulates resources for all public institutions. According to professional independence, the producers of official statistics shall decide independently and free from any pressures or interference for instance on the development, production, dissemination and communication of statistics, i.e. their

products. However, this development work could in practice be limited by not providing the necessary resources for the work, for instance in a case when the development of new statistics is not supported due to political reasons. In such a case the NSO should refer to its professional independence and try to justify the need for resources. At least, the Head of an NSO should be able to decide on the priorities based on user consultations when they allocate resources internally.

International collaboration

84. One of the important strengths of the current statistical systems is the strong international collaboration and joint work in an international statistical community. This work contributes to the development of international standards, increased comparability, harmonization and quality of statistics, stronger statistical capacity and higher professional independence. Based on assessments, peer reviews, audits and recommendations from international statistical organizations, internationally agreed standards contribute to achieving a high level of professional independence and strong public confidence, and help attract the technical assistance needed.

Access to microdata for researchers

85. Many NSOs provide access to microdata for researchers in strictly defined and regulated conditions. This enables opportunities to develop new insights into changes in society and to enhance the understanding of economic, social and environmental developments. Close cooperation with researchers may also help to identify improvement needs in statistical data and to develop new statistics and statistical services in joint projects with researchers.

Evolving role of national statistical offices – data stewardship

86. The digitalization of the economy and society has led to tremendous developments in what is referred to as the data ecosystem – the data, along with the data subjects, a broad range of stakeholders and data users, capacities, processes, policies and infrastructure used to capture and analyse data. Not only has the sheer amount of data that is being produced drastically increased, but government and other public agencies are increasingly becoming producers of data and data services beyond the production of official statistics. Globally, NSOs are discussing how best to position themselves to protect the integrity and relevance of official statistics, and - more generally - contribute to the treatment of data as a long-term asset that benefits society.

87. A highly relevant term for this changing role is data stewardship. As defined by a UNECE Task Force on this topic, data stewardship means *"ensuring the ethical and responsible creation, collection, management, use and reuse of data so that they are used for public good and benefit the full community of data users"* (UNECE, 2023a). According to the report by the Task Force, the NSO may not only steward the data produced and used within the NSO and be a system-wide data steward for the national statistical system but is also uniquely positioned to take on data stewardship responsibilities across the public sector. Especially the last of these three aspects of data stewardship expands the traditional role of NSOs. The use of data in the public sector goes well beyond the production of official statistics and is also concerned with new data services that government agencies provide.

88. NSOs can contribute to stewarding data across the public sector in many ways, for example by:

(a) Contributing to the development and promotion of data strategies, policy and principles;

(b) Coordinating the standardization and harmonization process and supporting interoperability;

(c) Providing expertise in data quality, and developing and promoting data quality frameworks;

(d) Contributing to improving data literacy;

(e) Providing expertise and services related to methods, tools and capabilities;

(f) Building partnerships across the public sector, secondments.

89. There are also other data services that many NSOs provide that fall under the umbrella of data stewardship, such as the provision of researcher access to observation-level data while ensuring confidentiality (for example using pseudonymized data).

2.2.2 Strengths of NSOs in the context of climate change-related information needs

90. There are several features of official statistical systems which can be emphasized due to their usefulness in climate change statistics (UNECE, 2014).

91. Official statistics adhere to international principles that bring a number of important attributes to the discourse on climate change. Official statistics are an important source of reliable information because of the strict conditions, professional independence and quality criteria under which they are produced. Other attributes include a commitment to the use of robust, transparent and commonly agreed methodologies, and a commitment to data accessibility. Generally, the national statistical office has a coordinating role in the national statistical system, while being part of international statistical systems that focus on the production of official statistics at regional or global level.

92. Official statistics often also include a link to the geographic region, and, in some countries, exact co-ordinates are attributed to statistics through geo-referencing. This valuable practice could be further developed as the spatial dimension is important in the context of climate change.

93. Because longitudinal studies are both time consuming and expensive, they are relatively rare in the published scientific literature. While cross-sectional scientific studies are essential in providing a detailed picture of a particular issue at a particular point in time, they do not provide a picture of how the issue is evolving over time. Official statistics, on the other hand, generally ensure a consistent time series, which can be of particular value in the examination of long-term trends.

94. The statistical system also has routines for developing statistics that are internationally comparable. This is done through a global institutional infrastructure to ensure that the definitions, classifications and data collection are harmonized across countries.

95. Furthermore, a core goal of the statistical system is the production of timely statistics. Official statisticians have developed their methods and data collection to improve timeliness, so statistical offices can help respond to the demand for timely climate change information.

96. Official statistics use well-developed methods for revising data, thanks to which changes are comparable over time, for example by taking into account the variability of factors affecting a given phenomenon. Long time series available in national statistical systems can be useful in revising GHG inventories and for climate change research.

97. In the context of assessing the socio-economic aspects of climate change, the statistical system may provide data to be linked to climate information in order to provide a broader and more informed description of the socio-economic aspects of climate change.

98. In recent years, the official statistics' focus on environmental and climate issues has broadened with many offices compiling environmental accounts measuring natural resource assets, energy, waste, water and air emissions as well as environmental protection expenditures, environmental taxes and subsidies showing the interaction between the environment and the economy. These accounts can support the development of statistics related to climate change. Further, as governments introduce various climate and environmental policies, these need to be classified and captured in macro-economic reporting.

2.3 Statistical frameworks and indicator sets

99. This section introduces existing statistical frameworks and indicator sets most relevant for climate change-related statistics for reference in other chapters.

2.3.1 Framework for the Development of Environment Statistics

100. The United Nations Framework for the Development of Environment Statistics (FDES) (United Nations, 2017) is a multipurpose conceptual and statistical framework that provides an organizing structure to guide the collection and compilation of environment statistics.

101. Due to its large coverage, FDES is a valuable tool providing a set of environmental topics and individual statistics to inform on climate change. FDES consists of six main components, each of which includes some climate-change related data:

(a) Environmental conditions and quality, which includes, for example, air temperature or precipitation;

(b) Environmental resources and their use, which includes land use change or non-renewable production;

(c) Residuals, which includes GHG released to the atmosphere;

(d) Extreme events and disasters, which includes the impact of droughts or floods;

(e) Human settlements and environmental health, which includes data on epidemic changes;

(f) Environmental protection, management and engagement, which includes cost of climate change mitigation and adaptation measures.

102. Section 5.3 of FDES presents a cross-cutting application of FDES to the issues of climate change.

103. In FDES, climate change-related statistics are organized in four blocks based on the sequence of climate change events used by the Intergovernmental Panel on Climate Change (IPCC):

(a) Climate process drivers;

(b) Climate change evidence;

(c) Climate change impacts and vulnerability;

(d) Mitigation and adaptation.

2.3.2 System of Environmental-Economic Accounting

104. The System of Environmental-Economic Accounting - Central Framework (SEEA-CF, United Nations et al., 2014)) has been developed collectively by the United Nations, the European Commission, FAO, IMF, OECD and the World Bank to integrate environmental and economic information in one common framework. In 2012 the United Nations Statistical Commission approved SEEA-CF as a statistical standard for environmental accounting and encouraged all countries to compile their environmental-economic accounts on the basis of it as well as to report statistics derived from it. The SEEA-CF contains internationally agreed concepts, definitions, classifications, accounting rules and tables for producing comparable statistics on the environment and its relationship with the economy. An important characteristic of environmental accounting is that the data are consistent with the System of National Accounts (SNA) principles, definitions and classifications.

105. SEEA-CF accounts can be used to monitor and analyse a wide scale of environmental issues, including climate change, and several SEEA-CF accounts can inform various aspects of climate change.

106. The SEEA-CF contains three main types of accounts

(a) Flow accounts: record the supply of resources - e.g. minerals, timber, fish - from the environment to the economy, the flows/use of products within the economy and the flows of residuals from the economy to the environment (e.g. solid waste and air emissions) in both physical and monetary terms;

(b) Asset accounts: record the stock of a specific environmental asset at the beginning and at the end of the accounting period and the changes (additions and reductions) during the accounting period; accounts can be in physical and monetary terms; examples are asset accounts for land;

(c) Environmental activity accounts and related flows: record monetary flows associated with economic activities whose primary purpose is environmental protection and resource management.

107. All three types of accounts can be used to analyse climate change-related issues. For example, physical flow accounts allow to identify GHG emissions by economic industry and households, which can be compared with economic information from the SNA. The accounts for monetary transactions include expenditure for actions and activities to reduce, prevent or eliminate GHG emissions. In addition, water asset accounts describe the changes in precipitation regimes and their implications for water stocks.

108. A point worth noting is the use of "residence principle" in the SEEA, which is different from the "territory principle" used in UNFCCC prescribed formats. The residence principle defines the scope of the national economy and thus what is included in the accounts. SEEA Air Emission Accounts use the residence principle (as in the national accounts) and record air emissions arising from activities of all resident units constituting a given national economy, regardless of where these emissions actually occur. It is important to note that national emission inventories required to be reported to the UNFCCC and the underlying activity data (e.g. energy statistics) do not follow the residence principle as applied in air emission accounts. National emission inventories rather follow a territory principle, i.e. they record emissions arising from the territory of a given country regardless of who is emitting (i.e. resident units or non-residents). Figure 2.8 illustrates the differences between the residence principle followed in the SEEA Air Emission Accounts (vertical ellipse) and the territory-oriented principle applied in national emission inventories (dotted horizontal ellipse). The latter count all emissions occurring on the territory originating from activities of resident units and non-residents. The residence principle calls for recording all emissions by resident unit on the territory but also outside, i.e. in the rest of the world.

Figure 2.8.
"Residence principle" in the Air Emission Accounts and "territory principle" in greenhouse gas inventories

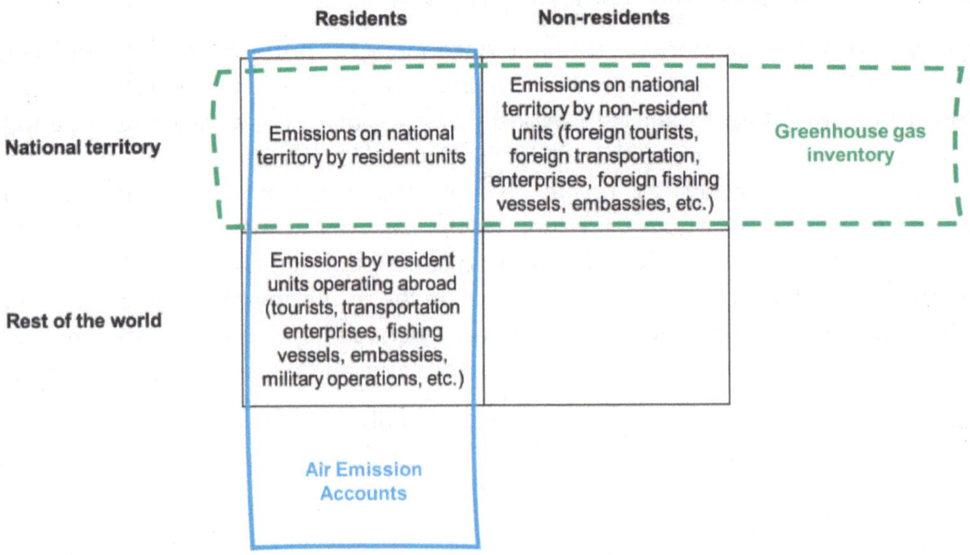

Source: Authors

109. The SEEA Ecosystem Accounting (SEEA-EA) complements the SEEA-CF by taking a different perspective (United Nations et al., 2021). The SEEA-EA takes the perspective of ecosystems and considers how environmental assets interact as part of natural processes within a given spatial area.

The SEEA-EA was adopted by the UN Statistical Commission in 2021 and provides a statistical framework to measure changes in the extent/size of ecosystems, their condition and the services they provide. The SEEA-EA contains carbon stock accounts and global climate regulation service accounts (e.g. accumulation and retention of carbon) which play an important role in complementing information on emissions. The SEEA-EA also contains several other accounts which can help measure the impacts of climate change on ecosystems and their ecosystem services (e.g. rainfall pattern regulation services, air filtration services, water supply, etc.).

2.3.3 International Recommendations for Energy Statistics

110. The International Recommendations for Energy Statistics (IRES) provide a comprehensive methodological framework for collecting, compiling and disseminating energy statistics in all countries irrespective of the level of development of their statistical system (United Nations, 2018). The Recommendations cover all aspects of the statistical production process, from the institutional and legal framework, basic concepts, definitions and classifications to data sources, data compilation strategies, energy balances, data quality issues and statistical dissemination. The Recommendations were developed at the request of the United Nations Statistical Commission, at its thirty-seventh session (7–10 March 2006) by UNSD in close cooperation with the Oslo Group on Energy Statistics and the Intersecretariat Working Group on Energy Statistics.

2.3.4 IPCC Guidelines for National GHG Inventories

111. The 2006 IPCC Guidelines for National Greenhouse Gas Inventories (hereafter 2006 IPCC Guidelines) provide methodologies for estimating national inventories of anthropogenic emissions by sources and removals by sinks of greenhouse gases (GHGs). The 2006 IPCC Guidelines were prepared in response to an invitation by the Parties to the UNFCCC, to assist Parties in fulfilling their commitments under the UNFCCC on reporting on inventories of anthropogenic emissions by sources and removals by sinks of GHGs not controlled by the Montreal Protocol, as agreed by the Parties.

112. The 2006 IPCC Guidelines are in five volumes. Volume 1 describes the basic steps in inventory development and offers the general guidance in GHG emissions and removals estimates based on the authors' understanding of accumulated experiences of countries over the period since the late 1980s, when national GHG inventories started to appear in significant numbers. Volumes 2 to 5 offer the guidance for estimates in different sectors of economy. The Guidelines are accompanied by an overview and a glossary.

Table 2.1.
Content of the 2006 IPCC Guidelines

Volumes	Chapters
1 - General Guidance and Reporting	1. Introduction to the 2006 Guidelines 2. Approaches to Data Collection 3. Uncertainties 4. Methodological Choice and Identification of Key Categories 5. Time Series Consistency 6. Quality Assurance/Quality Control and Verification 7. Precursors and Indirect Emissions 8. Reporting Guidance and Tables
2 - Energy	1. Introduction 2. Stationary Combustion 3. Mobile Combustion 4. Fugitive Emissions 5. CO2 Transport, Injection and Geological Storage 6. Reference Approach
3 - Industrial Processes and Product Use	1. Introduction 2. Mineral Industry Emissions 3. Chemical Industry Emissions

	4. Metal Industry Emissions 5. Non-Energy Products from Fuels and Solvent Use 6. Electronics Industry Emissions 7. Emissions of Fluorinated Substitutes for Ozone Depleting Substances 8. Other Product Manufacture and Use
4 - Agriculture, Forestry and Other Land Use	1. Introduction 2. Generic Methodologies Applicable to Multiple Land-use Categories 3. Consistent Representation of Lands 4. Forest Land 5. Cropland 6. Grassland 7. Wetlands 8. Settlements 9. Other Land 10. Emissions from Livestock and Manure Management 11. N_2O Emissions from Managed Soils, and CO_2 Emissions from Lime and Urea Application 12. Harvested Wood Products
5 - Waste	1. Introduction 2. Waste Generation, Composition and Management Data 3. Solid Waste Disposal 4. Biological Treatment of Solid Waste 5. Incineration and Open Burning of Waste 6. Wastewater Treatment and Discharge

Source: IPCC (2006).

113. In 2019, IPCC published 2019 Refinement to the 2006 IPCC Guidelines for National Greenhouse Gas Inventories (IPCC, 2019). The Refinement was adopted and accepted during the 49th Session of the IPCC in May 2019. It was developed to support the continuous improvement of national GHG inventories of emissions by sources and removals by sinks, taking into account scientific and other technical advances that had matured sufficiently since 2006.

2.3.5 Disaster-related Statistics Framework

114. The ESCAP Disaster-Related Statistics Framework (DRSF) was developed by the Expert Group on Disaster-related Statistics in Asia and the Pacific between 2014 and 2018 (ESCAP, 2018). The UNECE Task Force on measuring hazardous events and disasters contributed to the work and prepared the publication *CES Recommendations on the Role of Official Statistics in Measuring Hazardous Events and Disasters* (UNECE, 2019c).

115. DRSF is designed for use by national agencies to improve quality and harmonization of statistics in support of monitoring the Sendai Framework for Disaster Risk Reduction 2015-2030 and the Sustainable Development Goals. It provides the foundation for the development of a Global Framework on Disaster-related Statistics by the Inter-Agency and Expert Group on Disaster-related Statistics (IAEG-DRS), which was established under the umbrella of the UN Statistical Commission (50th Session, Decision 50/116) in 2020.

116. Figure 2.9 presents the main components of DRSF. These components are also closely linked with the indicator framework used for presenting the CES Set of Core Climate Change-related Indicators (organized in 5 areas: climate change drivers, GHG emissions, impacts, mitigation and adaptation).

Figure 2.9.
Disaster-related Statistics Framework

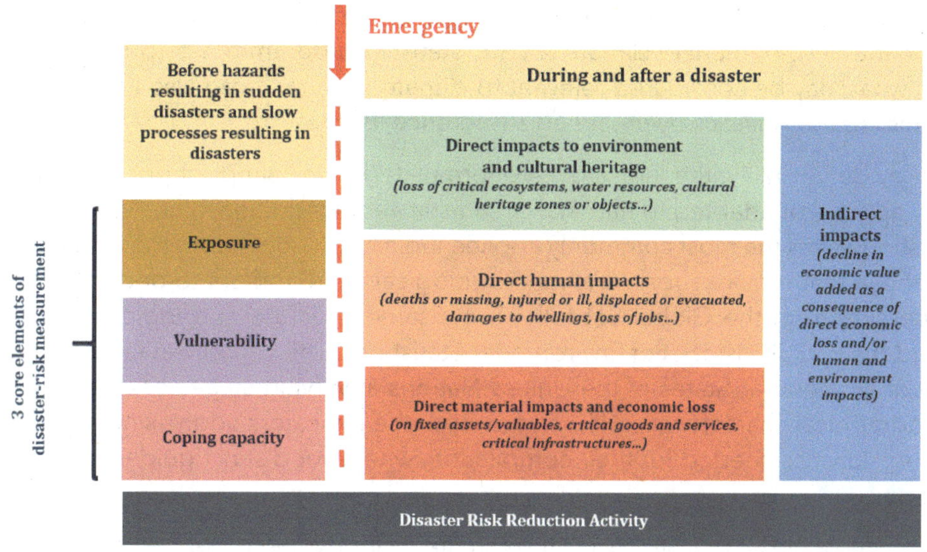

Source: ESCAP (2018).

2.3.6 CES Set of Core Climate Change-Related Statistics and Indicators

117. The CES Set of Core Climate Change-Related Indicators and Statistics (UNECE, 2021a) is a set of internationally comparable core climate change-related indicators and statistics, developed by a dedicated UNECE Task Force established by the Conference of European Statisticians. This indicator set was endorsed in 2020 by the Heads of NSOs of more than 60 CES member countries.

118. The CES Indicator Set provides the foundation for developing national sets of climate change-related indicators intended to show the big picture of climate change-related issues, address the most relevant current policy questions and help meet upcoming information needs. It includes 44 indicators covering climate change drivers, emissions, climate change impacts, mitigation and adaptation. Out of 44 indicators, 9 are either identical or related to SDG indicators and 4 – to Sendai Framework indicators. The selection of the indicators followed a careful vetting process based on the criteria of policy relevance, methodological soundness and data availability described in the present document. The CES Indicator Set also proposes corresponding contextual and operational indicators, helping interpret the core set in the national and global context and provide more details according to national circumstances and priorities, and is accompanied by Implementation Guidelines (UNECE, 2021b). It is the recommended set of climate change-related indicators to be compiled and published in the CES member countries. The CES Indicator Set is complementary to the Global Set of Climate Change Statistics and Indicators (see the section below), which includes 33 out of 44 indicators from the CES Set, either verbatim or in a related form as it focuses on core indicators needed to paint the big picture of climate change in an internationally comparable way and respond to main policy questions relevant in the UNECE region.

119. The full CES Indicator Set is presented in Annex 3 and available on the UNECE webpage.[10]

2.3.7 Global Set of Climate Change Statistics and Indicators

120. The Global Set of Climate Change Statistics and Indicators (UNSD, 2022) was adopted at the fifty-third session of the United Nations Statistical Commission in March 2022 as the framework for climate change statistics and indicators.

[10] https://unece.org/statistics/ces-set-core-climate-change-related-indicators-and-statistics-using-seea

121. The Global Set is a comprehensive statistical framework, with statistics, indicators and metadata, designed to support countries in preparing their own sets of climate change statistics and indicators according to their individual concerns, priorities and resources. It will assist countries embarking on the development of climate change statistics programmes by providing the scope and coverage on what may be considered relevant to climate change. It can also assist countries already involved in this area of statistics by providing a reference list.

122. To provide flexibility, the indicators are formulated in the simplest possible way, so that they can be easily applied or adapted to national circumstances. In addition, a tiered system was set up in a way that distinguishes the most commonly applied indicators (Tier 1) from those that are less applied (Tier 2) and those that require substantial methodological development to become operational (Tier 3). The tiered system of the Global Set follows the same structure and logic applied for the SDG indicators and the FDES' Basic Set of Environment Statistics[11]. The Global Set takes into consideration the diversity of all Member States of the United Nations at varying stages of development and with different geographical characteristics. It includes both the indicators addressing climate change and the underlying statistics needed for their compilation, whenever distinct guidance for these statistics was identified.

123. The scope of the Global Set covers the climate change aspects defined by the five policy areas of the IPCC, namely drivers, impacts, vulnerability, mitigation and adaptation. The Global Set includes primarily the biophysical indicators and statistics, but also human activities, and social and institutional aspects related to climate change. The links between policy and statistics are articulated according to the relevant articles of the Paris Agreement and the subsequent decisions of CMA, as well as related SDGs and Sendai Framework indicators, and the FDES statistics. In total, the Global Set includes 158 indicators and 190 statistics. Out of those, 54 are related, similar or identical to SDG indicators and 14 - to the Sendai Framework indicators.

124. All statistics and indicators included in the Set and additional resources mentioned above can be found on the Global Set web pages[12]. The list of statistics and indicators included in the Global Set is also presented in Annex 4.

2.3.8 CES Set of Core Disaster-Risk-Related Indicators

125. The CES Set of Core Disaster-Risk-Related Indicators (UNECE, 2023b) was developed by a dedicated CES Task Force on measuring hazardous events and disasters to support the implementation of the *CES Recommendations on the Role of Official Statistics in Measuring Hazardous Events and Disasters* and endorsed by the Conference of European Statisticians in June 2023.

126. The proposed set of core disaster-risk related indicators focuses on measuring disaster risk, which is at the heart of disaster-risk management. Many of its elements can be measured with existing statistics. These indicators are recommended for regular production by NSOs as they (a) strengthen evidence for disaster risk; (b) inform about the state of disaster risk in an internationally comparable way; (c) support monitoring and reporting on international policy agreements; (d) ensure consistency and coherence of information across administrative boundaries; and (e) add value to existing statistics. The proposed indicator set complements the *CES Set of Core Climate Change-Related Indicators and Statistics Using SEEA*.

2.3.9 Sendai Framework indicators

127. The Sendai Framework Indicators consist of 38 indicators, to measure progress in achieving the global targets of the Sendai Framework and determine global trends in the reduction of risk and losses. Six of the 38 indicators are so called compound indicators of some or all the indicators for a global target.

[11] https://unstats.un.org/unsd/envstats/fdes/basicset.cshtml
[12] https://unstats.un.org/unsd/envstats/climatechange.cshtml

128. The Sendai Framework Indicators were developed by the open-ended intergovernmental expert working group on indicators and terminology relating to disaster risk reduction, which was established by the General Assembly during its 69th session. The group's proposal for the Sendai Framework Indicators was endorsed by the General Assembly in February 2017.

129. The working group considered the collection of disaggregated data for the indicators instrumental to the effective implementation of the Sendai Framework and relevant disaster risk related targets of the Sustainable Development Goals (United Nations General Assembly, 2016).

2.3.10 SDG Global Indicator Framework

130. The global indicator framework for the Sustainable Development Goals and Targets of the 2030 Agenda for Sustainable Development is a large multi-domain indicator framework with - as of 2023 - 231 unique indicators, which often include multiple statistical series. The internationally agreed metadata also prescribes how series should be disaggregated, for example by sex, age or geographically.

131. As of March 2023, the methodology and standards have been established for all indicators. However, issues around data availability remain - 77 indicators are classified as Tier II: Indicator is conceptually clear, has an internationally established methodology and standards are available, but data are not regularly produced by countries.[13]

132. Important synergies exist between the SDG indicators and the Sendai Framework Indicators. The 5 compound indicators out of 7 Global Targets of the Sendai Framework, are included in the SDG indicator framework and the monitoring of the Sendai Framework is intended to complement the relevant SDG indicators.[14]

133. Relevant indicators from the SDG global indicator framework are included in the Global Set of Climate Change Statistics and Indicators, the CES Set of Core Climate Change-Related Indicators and the Set of Core Disaster-Risk-Related Indicators.

134. The Indicators for the Kunming - Montreal Global Biodiversity Framework, which have been adopted by the Parties to the Convention on Biological Diversity in December 2022 - is also including relevant SDG indicators as headline indicators for some of the goals and targets.

135. Noteworthy is the clearly established role of the international statistical community that is ingrained in the 2030 Agenda for Sustainable Development. The principal General Assembly resolution on the 2030 Agenda (United Nations, 2015) gave a mandate to the Statistical Commission to agree on a framework developed by an expert group consisting of representatives of the statistical offices of international agencies and experts from member states - the Inter-Agency and Expert Group on Sustainable Development Goal Indicators.[15] The General Assembly Resolution adopting the indicator framework also gives NSOs a strong mandate to be the coordinator of the National Statistical System in the context of SDG monitoring (United Nations General Assembly, 2017).

---

[13] See the Tier Classification for Global SDG Indicators as of 6 March 2024, https://unstats.un.org/sdgs/iaeg-sdgs/tier-classification/

[14] See more about the integrated monitoring of the Sendai Framework global targets and SDGs at https://www.preventionweb.net/sendai-framework/Integrated%20monitoring%20of%20the%20global%20targets%20of%20the%20Sendai%20Framework%20and%20the%20Sustainable%20Development%20Goals.

[15] See https://unstats.un.org/sdgs/iaeg-sdgs/

Chapter 3 Reporting under the Paris Agreement

3.1 Introduction

136. The chapter discusses the role of NSOs in addressing the reporting requirements under the Enhanced Transparency Framework (ETF) as adopted by the first Conference of the Parties serving as the meeting of the Parties to the Paris Agreement (CMA); held in Katowice in December 2018. These requirements include not just information on country's national GHG emissions and removals, but also the targets (nationally determined contributions, NDCs) progress in implementing and achieving their climate plans, financial resources, technology development and transfer as well as capacity building provided and received.

137. The reporting requirements, related data needs and steps in a typical reporting cycle are described in section 3.2. Section 3.3 discusses how NSOs can contribute to this process, while section 3.4 presents conclusions and recommendations.

3.2 Policy context and definitions

3.2.1 Reporting requirements related to UNFCCC under the Paris Agreement

138. As the Paris Agreement is built around individual climate plans and targets from all governments - nationally determined contributions - the tracking and reporting their progress towards the implementation and achievement of these plans will be conducted under the ETF. Every five years, governments will come together to participate in the global stocktake (GST) - a process of periodically taking stock of the implementation of the Paris Agreement to assess the collective progress towards achieving the purpose of the Agreement and its long-term goals.

139. Transparency of climate action and support, through the ETF, is viewed as an important part of the Paris Agreement (see section 2.1.1.1). For many, the ETF is a key mechanism for building mutual trust and confidence, while others additionally see the ETF as a tool for facilitating implementation and raising ambition of climate action. In contrast to the current Measurement, Reporting and Verification (MRV) arrangements under the Convention and the Kyoto Protocol (also discussed in more detail in section 2.1.1.1), the ETF under the Paris Agreement applies to all Parties through a common set of operational details - the ETF modalities, procedures and guidelines (MPGs). The Paris Agreement recognizes the national circumstances and capacity constraints of developing country Parties and as a result provides flexibility to developing country Parties that need it in the light of their capacities.

140. The ETF will serve as the primary vehicle for the governments to demonstrate progress in implementation and achievement of their NDCs which need to be prepared and communicated every five years. To facilitate clarity, transparency and understanding (ICTU) of NDCs, they need to include certain information elements for which detailed guidance exists, including inter alia, quantifiable information on the reference point of the NDC (see para 142(c)). Under the ETF, it is mandatory for all governments to report in biennial transparency reports (BTRs) on climate action and support, undergo a technical expert review of their BTRs as well as a facilitative multilateral consideration of progress. Parties are to submit the first BTR under the ETF no later than 31 December 2024 and subsequent reports every two years thereafter. The information from BTRs, including on tracking progress of the implementation and achievement of NDCs through a set of nationally defined indicators, from all Parties will feed into the GST, taking place first in 2023 and then every five years thereafter. The outcomes of the GST will in turn inform the preparation of more ambitious successive NDCs by all

Parties. Hence, the robustness of the GST depends to a large extent also on the quality of the inputs, including from BTRs and NDCs[16].

Figure 3.1
Outline of the core elements of the Enhanced Transparency Framework under the Paris Agreement

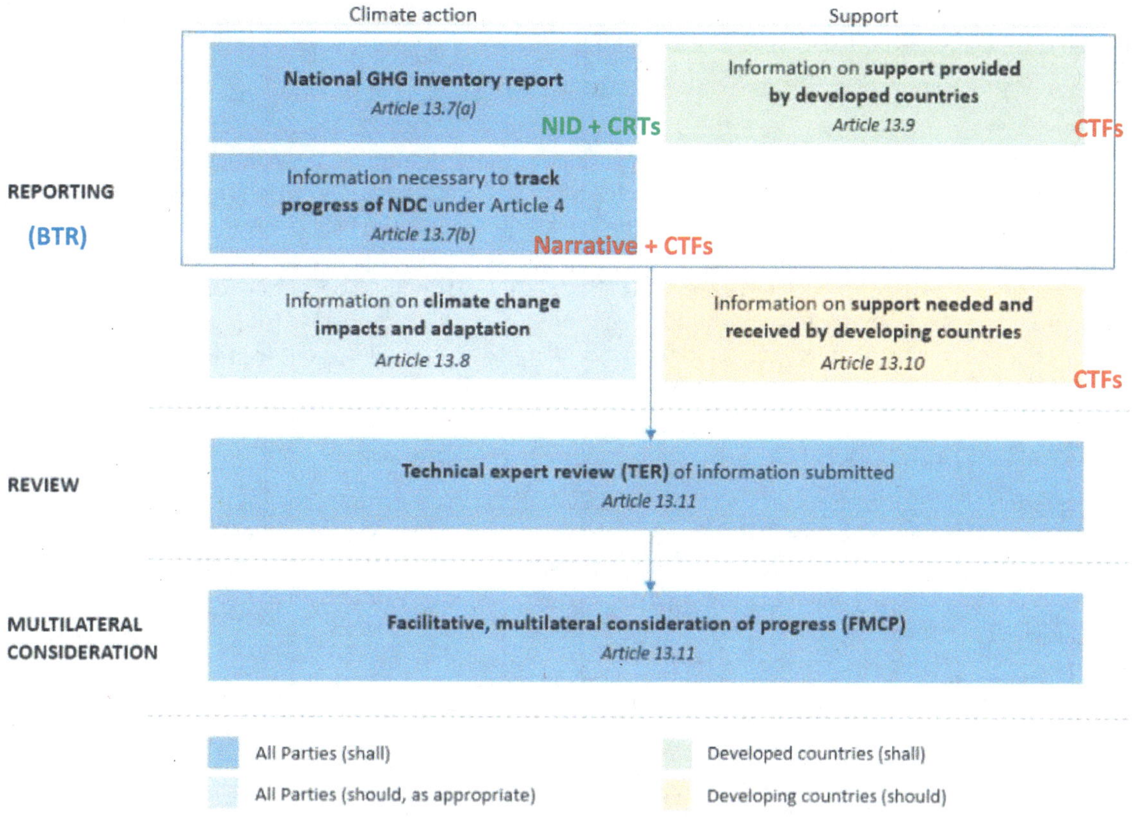

GLOSSARY

BTR = Biennial Transparency Report
NID = National Inventory Document
NDC = Nationally Determined Contributions
CRTs = Common Reporting Tables
CTFs = Common Tabular Formats

Source: Based on UNFCCC (2023), modified by the authors.

3.2.1.1 Nationally determined contributions

141. Under Article 4 of the Paris Agreement, Parties are required to communicate, every five years, their NDCs, which include countries' climate targets and plans to reduce emissions and may also describe how countries are planning to adapt to climate impacts.

142. As the name suggests, the scope of NDCs is self-determined by each Party, including the mitigation target and should represent a progression from the previous one. NDCs should include the following information that facilitates clarity, transparency and understanding (ICTU):

(a) Scope and coverage of the NDC (description of the target and the categories and gases covered);

(b) Time frame and/or periods for implementation (start and end date of the target);

(c) Quantifiable information on the reference point (information on reference indicator for the mitigation target, base year used, starting and/or target value or planned deviation from the starting value, and source data used);

[16] Further information on the information relevant for NDCs is contained in annex I and guidance for ICTU is available at: https://unfccc.int/sites/default/files/resource/cma2018_3_add1_advance.pdf#page=6

(d) Assumptions and methodological approaches;

(e) National planning processes for developing the NDC, and if available implementation plans, taking into account national circumstances;

(f) How the Party determines that its NDC is fair and ambitious;

(g) How the NDC contributes towards achieving the objective of the Convention.

Box 3.1. Understanding the nationally determined contributions

Nationally determined contributions (climate plans and targets) determine how much each country intends to reduce domestic emissions and adapt to the effects of climate change. They are an integral part of the commitments that parties must fulfil under the Paris Agreement. The NDCs cover all Parties to the UNFCCC and their climate plans, unlike the Kyoto Protocol to the UNFCCC, where the GHG emission reduction targets covered only selected countries. The NDCs cover mitigation targets or mitigation co-benefits resulting from adaptation actions and/or economic diversification plans. The mitigation targets range from economy-wide absolute emission reduction targets to strategies, policies, plans and actions for low-emission development. According to Art. 4 sec. 12 of the Paris Agreement, NDCs communicated by Parties are recorded in a public register maintained by the secretariat, available under: https://unfccc.int/NDCREG.

According to the UNFCCC Secretariat's report (UNFCCC, 2022a)., summarizing the information from the 166 available NDCs provided by the 193 Parties* to the Paris Agreement and registered in the Nationally Determined Contributions (as of 23 September 2022) shows that NDCs cover 94.9% of the total global emissions in 2019, which are estimated at 52.6 Gt CO_2 eq. without Land Use Land-Use Change and Forestry (LULUCF). In their NDCs:

- Most Parties (90%) provided quantified mitigation targets, expressed as clear numerical targets, while the rest (10%) included strategies, policies, plans and actions for which there is no quantifiable information as components of their NDCs;
- Most Parties (80%) communicated economy-wide targets, covering all or almost all sectors defined in the 2006 IPCC Guidelines, with an increasing number of Parties moving to absolute emission reduction targets in their new or updated NDCs;
- In terms of GHGs, all NDCs cover CO_2 emissions, most cover CH_4 (91%) and N_2O (89%) emissions, many (53%) cover HFC emissions and some cover PFC and SF_6 (36%) and NF_3 (26%) emissions;
- Many Parties (50%) provided information on mitigation co-benefits resulting from their adaptation actions and/or economic diversification plans, mostly in combination with other targets;
- Most of the Parties (74%) that submitted new or updated NDCs have strengthened their commitment to reducing or limiting GHG emissions by 2025 and/or 2030, demonstrating increased ambition in addressing climate change.

The projected total global GHG emission level taking into account full implementation of all latest NDCs (including all conditional elements) implies an even stronger possibility of global emissions peaking before 2030 than estimated in the previous version of the NDC synthesis report, with the lower bound of the 2030 emission level (49.1 Gt CO_2 eq.) estimated to be up to 6.6% below the 2019 emission level (52.6 Gt CO_2 eq.) and 5.1% below the lower bound of the estimated 2025 emission level (51.8 Gt CO_2 eq.). However, in order to achieve that peaking, the conditional elements of the NDCs need to be implemented, which depends mostly on access to enhanced financial resources, technology transfer and technical cooperation, and capacity-building support; availability of market-based mechanisms; and absorptive capacity of forests and other ecosystems.

* The difference between the number of NDCs and Parties is because the EU member States submit one NDC jointly.

3.2.1.2 Long-term low greenhouse gas emission development strategies

143. Article 4 of the Paris Agreement states that "*all Parties should strive to formulate and communicate long-term low GHG emission development strategies*". Long-term low GHG emission

development strategies (LT-LEDS), although not mandatory, facilitate long-term planning and identification of development priorities towards a country's future development. These strategies should be updated regularly in line with the best available science.

3.2.1.3 Enhanced Transparency Framework

144. Through the enhanced transparency framework (ETF), starting from 2024, countries will report on actions taken and progress in climate change mitigation, adaptation measures and support provided or received. This framework is applicable to all Parties and builds upon the existing MRV system under the Convention.

145. The elements of the ETF process include:

(a) Biennial transparency report (BTR) (see the following sections);

(b) Technical expert review (TER);

(c) Facilitative multilateral consideration of progress (FMCP).

146. The BTR and NDC are connected. Parties shall include in the BTR a description of their NDC. Also, the progress made by a Party in implementing and/or achieving its NDC will be reported in the "structured summary" of the BTR.

147. Article 13 of the Paris Agreement establishes the two clear purposes of the ETF, one on climate action and another on support for such action:

(a) With respect to climate action, the purpose of the ETF is "*to provide a clear understanding of climate change action in the light of the objective of the Convention as set out in its Article 2, including clarity and tracking of progress towards achieving Parties' individual NDCs under Article 4 of the Paris Agreement (herein after referred to as "NDCs)*", and Parties' adaptation actions under Article 7, including good practices, priorities, needs and gaps, to inform the global stocktake under Article 14;

(b) Similarly, with respect to support for climate action, the purpose of the ETF is "*to provide clarity on support provided and received by relevant individual Parties in the context of climate change actions under Articles 4, 7, 9, 10 and 11, and, to the extent possible, to provide a full overview of aggregate financial support provided, to inform the global stocktake under Article 14*".

148. The information gathered through the ETF will feed into the Global Stocktake (which is a 5-year cycle), through which the collective progress towards the long-term climate goals will be assessed. The assessment of the progress may lead to recommendations for ramping up ambitions by countries.

149. The reporting requirements under the ETF are operationalized through a set of modalities, procedures and guidelines (MPGs) for the transparency framework for action and support referred to in Article 13 of the Paris Agreement. MPGs were adopted by the first CMA; held in Katowice in December 2018[17] and are contained in decision 18/CMA.1 and its annex (UNFCCC, 2018).

150. To promote universal participation, the ETF and its MPGs include built-in flexibility which takes into account Parties' different capacities and build upon the collective experience of both developed and developing countries. The MPGs specify the flexibility provisions available to those developing country Parties that need it in the light of their capacities, pursuant to Article 13, paragraph 2, reflecting flexibility, including in the scope, frequency and level of detail of reporting, and the scope of the review

[17] See section 2.1.1.1 for description of the bodies under UNFCCC and the Paris Agreement

3.2.1.4 Biennial transparency reports

151. In reporting their BTRs, countries should include the following:

(a) A national inventory report of anthropogenic emissions by sources and removals by sinks of GHGs - mandatory for all Parties (under UNFCCC, for developed country Parties it is an annual obligation);

(b) Information necessary to track progress in implementing and achieving its NDC under Article 4 of the Paris Agreement - mandatory for all Parties;

(c) Information on climate change impacts and adaptation under Article 7 of the Paris Agreement - not mandatory;

(d) Information on financial, technology transfer and capacity-building support provided to developing country Parties - mandatory for developed country Parties;

(e) Information on financial, technology transfer and capacity-building support needed and received - reporting element for developing country Parties, but not mandatory.

152. As already mentioned, there is a relationship between the NDC and the ETF as Parties need to provide a description of their NDC in their BTR against which progress made will be tracked (decision 18/CMA.1, para. 64). The information on a Party's NDC that is required to be provided in the BTR is similar, but not identical, to the information that is necessary for transparency, clarity and understanding in their NDC (decision 4/CMA.1, para. 7).

153. The above information should be reported in a narrative format, as well as in the agreed Common Reporting Tables (CRT) or Common Tabular Formats (CTF), as appropriate (decision 5/CMA.3 in UNFCCC (2021b)). It is however important to note that some flexibility is provided to developing country Parties that need it in the light of their capacities in terms of the scope, frequency and level of detail of reporting.

3.2.2 Data related to the reporting requirements

154. A brief description of the information to be provided, in particular those relating to statistics and indicators, is presented below. Further information in understanding the indicators relevant for these reporting requirements under the ETF and its linkages to other processes under the UNFCCC, are provided in the Reference Manual for the Enhanced Transparency Framework under the Paris Agreement (UNFCCC, 2022b).

155. Common Reporting Tables/Common Tabular Formats have been prescribed for the following information:

(a) Information in the national inventory reports of anthropogenic emissions by sources and removals by sinks of GHGs;

(b) Information necessary to track progress made in implementing and achieving nationally determined contributions under Article 4 of the Paris Agreement; and

(c) Information on financial, technology development and transfer and capacity-building support provided and mobilized, as well as support needed and received, under Articles 9–11 of the Paris Agreement.

156. Further details are discussed in the following sections.

3.2.2.1 *National inventory report of anthropogenic emissions by sources and removals by sinks of GHGs*

157. The national inventory report (NIR) is comprised of a national inventory document (NID) and agreed common reporting tables.[18] Parties to the Paris Agreement shall follow the methodology recommended in the 2006 IPCC Guidelines for National Greenhouse Gas Inventories to estimate their anthropogenic emissions by sources and removals by sinks, and use the 100-year time-horizon global warming potential values from the IPCC Fifth Assessment Report to present aggregate emissions in CO_2 equivalent.

158. Each Party shall report the following sectors: energy; industrial processes and product use; agriculture; land use, land-use change and forestry (LULUCF); and waste. Data on international aviation and marine bunker fuel emissions should also be reported but excluded from the national total emissions.

159. In preparing their national inventory report, all Parties shall use the 2006 IPCC Guidelines, and any subsequent version of or refinement to these IPCC guidelines agreed upon by the CMA. Further, each Party is encouraged to use the Wetlands Supplement.[19]

160. The MPGs also provide that a Party should use nationally appropriate methodologies if they better reflect its national circumstances and are consistent with the IPCC guidelines. For such cases, the Party must transparently explain the national methods, data and/or parameters selected.

161. Each Party should make every effort to use a recommended method (tier level) for key categories. Parties are encouraged to apply the highest tier level where possible. A Party may be unable to apply a higher tier method for a particular key category owing to a lack of resources. In such cases, the Party may use a tier 1 approach, and shall clearly document why the methodology used was not in line with the corresponding decision tree of the IPCC guidelines. The Party should prioritize for future improvement any key categories for which the good practice method elaborated in the IPCC guidelines cannot be used.

162. Parties are encouraged to use country-specific and regional emission factors and activity data, where available, or to propose plans to develop such emission factors and activity data in accordance with the IPCC guidelines.

163. Parties must quantitatively estimate and qualitatively discuss the uncertainty of the emission and removal estimates for all source and sink categories, including inventory totals, for at least the starting year and the latest reporting year of the inventory time series. It is also essential to estimate the trend uncertainty of emission and removal estimates for all source and sink categories, including totals, between the starting year and the latest reporting year of the inventory time series, using at least approach 1 contained in the 2006 IPCC guidelines. Those developing country Parties that need flexibility in the light of their capacities may provide, at a minimum, a qualitative discussion of uncertainty for key categories in the event that input data are unavailable to quantitatively estimate uncertainties. At the same time, these developing country Parties are encouraged to provide quantitative uncertainty estimates for all source and sink categories of the GHG inventory.

164. The estimates of emissions and removals shall be reported in units of mass and in CO_2 for the following gases: carbon dioxide (CO_2), methane (CH_4), nitrous oxide (N_2O), hydrofluorocarbons (HFCs), perfluorocarbons (PFCs), sulphur hexafluoride (SF_6) and nitrogen trifluoride (NF_3). Data for HFCs and PFCs shall be disaggregated by chemical, such as HFC-23. Moreover, data should also be provided for the following precursor gases: carbon monoxide (CO), nitrogen oxides and non-methane volatile organic compounds (NMVOCs), as well as sulphur oxides.

165. Data required for compiling the inventories includes the emission factors and activity data at the most disaggregated level, and these data is also required to be reported, to the extent possible.

---

[18] Available at https://unfccc.int/documents/311076.

[19] In this chapter, the term "the IPCC guidelines" refers to the 2006 IPCC Guidelines and the Wetlands Supplement together.

166. The above-mentioned information shall be reported for a time series covering the period from 1990 until the latest reporting year, which shall be no more than two years prior to the submission of its national inventory report. For example, the submission in 2024 shall cover the period 1990–2022.

3.2.2.2 Information necessary to track progress in implementing and achieving its NDC

167. This information shall be reported in narrative format in the relevant section of the BTR, together with the structured summary of information relating to indicators presented in the agreed CTF (annex II in UNFCCC, 2021b).

168. In addition to other information that a Party may deem useful to tracking progress made in implementing and achieving its NDC under Article 4 of the Paris Agreement, below is the relevant information to be reported.

National circumstances

169. Each Party shall report on its national circumstances that are relevant to the progress made in implementing and achieving its NDC under Article 4 of the Paris Agreement. Such information can include the country's population, economic and climate profile.

Tracking progress

170. In order to track the progress made in implementing and achieving its NDC, each Party shall report on qualitative or quantitative indicators. Such indicators could include net GHG emissions and removals, percentage reduction of GHG intensity, qualitative indicators for a policy or measure, hectares of reforestation, percentage of renewable energy use or production, share of non-fossil fuel in primary energy consumption and other non-GHG related indicators.

171. It is important to note that each Party shall identify and select the indicators that it will use to track progress towards its target indicated in the NDC. Most recent information shall be reported for each selected indicator each reporting year during the implementation period of its NDC.[20]

172. Each Party shall also provide information on the use of internationally transferred mitigation outcomes towards an NDC if it voluntarily participates in cooperation mechanisms under Article 6 of the Paris Agreement.

Mitigation policies and measures, actions and plans

173. Information on actions, policies and measures shall be organized, to the extent possible, by the GHG inventory sectors, including international transport.

174. For the individual actions, policies and measures, the costs, non-GHG mitigation benefits, and estimates of expected and achieved GHG emission reductions should be reported.

Projections of greenhouse gas emissions and removals

175. Each Party shall report projections by sector and by gas, including total GHG emissions with and without emissions from the LULUCF sector, using the same metric consistent with its national GHG inventory. Projections shall be provided for at least 15 years beyond the latest year of the national GHG inventory ending in zero or five.

176. The key underlying assumptions and parameters used in developing projections, such as population and economic growth, should be reported.

3.2.2.3 Information on climate change impacts and adaptation

177. If a Party wishes to do so, the information listed below, among others, could be provided to recognize its adaptation efforts:

[20] To complement the tracking process using different indicators per Party under the Paris Agreement, OECD makes a set of harmonized indicators on GHG emission trends and targets available on the dashboard of the International Programme for Action on Climate (IPAC) – Pizarro R., et al. (2024).

(a) National circumstances relating to adaptation, such as demographics, economy, infrastructure and adaptive capacity;

(b) Current and projected climate trends and hazards;

(c) Observed and potential impacts of climate change, including sectoral, economic, social and/or environmental vulnerabilities;

(d) Adaptation challenges and gaps, and barriers to adaptation;

(e) Adaptation strategies, policies, plans, goals and actions, including progress on their implementation;

(f) Approaches and systems for monitoring and evaluation, including assessment of and indicators for;

(i) How adaptation increased resilience and reduced impacts;

(ii) Effective implemented adaptation measures;

(g) Information related to averting, minimizing and addressing loss and damage associated with climate change impacts.

3.2.2.4 Information on financial, technology transfer and capacity-building support provided and mobilized

178. Developed country Parties shall provide this information in narrative format in the relevant section of the BTR, together with quantitative/qualitative information presented in the agreed CTF tables (annex III to UNFCCC, 2021b).

3.2.2.5 Information on financial, technology transfer and capacity-building support needed and received

179. Developing country Parties should provide this information in narrative format in the relevant section of the BTR, together with quantitative/qualitative information presented in the agreed CTF tables.

3.2.3 Steps in a typical reporting cycle

180. Meeting the reporting obligations under the Paris Agreement is a process, and the establishment of robust and sustainable institutional arrangements is a key enabling factor. This includes building capacity and expertise within a country so that necessary information can be collected and provided in a timely and reliable manner.

181. Parties to the Paris Agreement have to report GHG inventories according to the common MPGs of the Enhanced Transparency Framework (ETF). This is a mandatory requirement for all Parties, with embedded flexibilities for developing country Parties that need them in light of their capacities.

182. In particular, each Party shall report a national GHG inventory under the Paris Agreement every two years using the common reporting formats agreed at COP26 in Glasgow. In relation to methodological choice, countries shall use the 2006 IPCC Guidelines and/or, on a voluntary basis, the 2019 IPCC Refinement to the 2006 IPCC Guidelines (IPCC, 2019).

183. Annex I Parties to the Convention have been reporting GHG inventories at a level of detail similar to the common tables agreed for reporting under the Paris Agreement. For non-Annex I Parties, the reporting of GHG inventories had been linked to National Communications and Biennial Update Reports, where the level of detail in the reporting is substantially less detailed. (See section 2.1.1.1 for more information on Annex I, Annex II and non-Annex I Parties)

184. Although some developing countries have good inventories, there will be a steeper learning curve for most of them, particularly for those countries where inventory systems remain a challenge.

However, the quality of the first inventories reported by Annex I Parties about 20 years ago was also far below the quality of their inventories today. A GHG inventory takes time to develop and improve, and progress can be slow, but good institutional arrangements should set the basis for sustained quality improvements over time.

185. In relation to quality, the IPCC promotes five key concepts to assess the quality of GHG inventories: transparency, accuracy, consistency, comparability and completeness.

Figure 3.2.
Steps in a typical inventory cycle

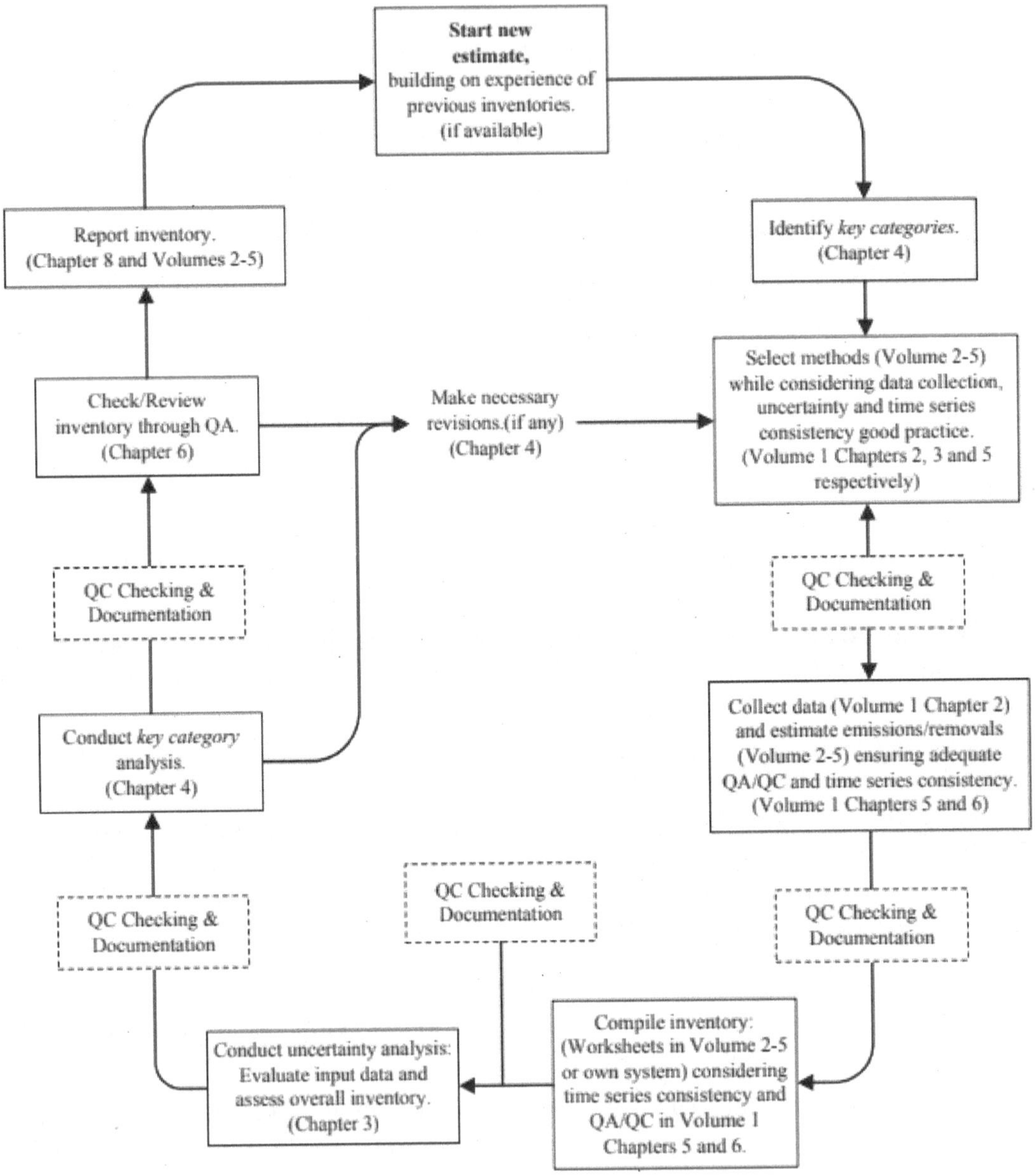

Source: IPCC (2006).

186. The IPCC Guidelines also provide information about the typical steps in the GHG inventory cycle, which can be useful to countries to compile a revised or new GHG inventory and reporting. The first thing to highlight is that the inventory compiler should always base any improvements and future revisions of the inventory on previous inventories, in an iterative process. Figure 3.2 illustrates the steps in a typical inventory cycle according to the 2006 IPCC Guidelines. To summarize the figure, the inventory compilation process has the following steps: 1) identify the key categories (those most important in terms of emissions in the country); 2) identify the appropriate estimation method using the IPCC Good Practice decision trees; 3) collect the necessary data; 4) estimate emissions and

removals using the selected methods and the data collected; 5) carry out the uncertainty analysis; 6) conduct the key category analysis; 7) do the Quality Assurance of the whole inventory; and, 8) report the inventory. The Quality Control procedures should be implemented and documented for each of these steps, to facilitate the review of the inventory.

187. The objective is to present an inventory that is transparent and comparable, and as clear as possible to enhance credibility and enable users to understand the data, methods and assumptions used in the estimation of emissions and removals.

188. Inventory compilers should base future inventories on previous inventories in an iterative process that improves the inventory on a continuous basis. When the new inventory is compiled, estimates for all years should be reviewed and updated, on the basis of any improvements resulting from the implementation of recommendations from GHG inventory reviews, audits, or own improvements.

189. Figure 3.3 shows an overview of the flexibility provisions in the MPGs, and where they can apply. The column with the 'shall' provisions is what developed country Parties have to report. The column with 'flexibilities applicable' is what developing country Parties may report in light of their capacities, for a limited period of time. The MPGs require these Parties to clarify any capacity constraints when applying specific flexibilities and provide estimated time frames for improvements. Specific information on the potential use of the flexibility provisions is included in Decision 18/CMA.1 (MPGs).

Figure 3.3.
Overview of the flexibility provisions in the modalities, procedures, and guidelines (MPGs) of the Enhanced Transparency Framework

Reporting elements	"Shall" provision in MPGs	Flexibilities applicable
Key category analysis	95% threshold	85% to 95% threshold
Uncertainty assessment	Quantitative and qualitative for all categories	At least qualitative for key categories
Insignificance threshold	'NE' f lower than 0.05% of NT and 500 Kt of CO_2e	'NE' if lower than 0.1% of NT and 1000 Kt of CO_2e
QA/QC	Develop QA/QC plan	Encouraged to develop
GHGs	CO_2, CH_4, N_2O, HFCs, PFCs, SF_6, and NF_3	CO_2 CH_4, N_2O. Other if n NDC or reported before
Time series	Annually between 1990 and t-2	Reference year for NDC and annually from 2020 to t-3

***Source*: Adapted from decision 18/CMA.1 (MPGs); UNFCCC (2018).**

190. It is clear that Parties are at different levels of reporting, and this is recognized in both the IPCC Guidelines, when estimating emissions and removals, and in the MPGs, in terms of reporting. Some of the key 'guiding principles' from the MPGs include the importance of facilitating improved reporting and transparency over time; providing flexibility to those developing country Parties that need it in the light of their capacities; and promoting transparency, accuracy, completeness, consistency, and comparability. The regular reporting and review of GHG inventories, consistent with the MPGs and the IPCC Guidelines, should assist countries in improving their inventories as well as effective implementation of these 'guiding principles'.

3.3 How NSOs can contribute

191. The role of NSOs in GHG inventories is not only important and has been strengthened in recent years but is also likely to increase due to the needs arising from the Paris Agreement implementation. This is particularly important for developing country Parties where the right data to estimate emissions and removals, not only for GHG inventories but also for indicators to track progress towards countries' climate change commitments are often not available.

192. One possibility is to include NSOs in the national GHG inventory system. In such a situation, it seems necessary to formalize institutional relations, which would provide certainty regarding the roles and responsibilities of each of the institutions involved. This decision should be made by the competent authorities in the country, as there are advantages to NSO participation in the formal inventory process, as NSOs can help to ensure and improve the quality of GHG inventories in a sustainable way. Various institutional arrangements are also possible and discussed in this section.

193. Currently, the level and formalities of NSO involvement in GHG inventories varies among countries - from GHG inventory unit being part of the NSO (Finland, Türkiye) or responsible for part of the inventory (Norway) through NSOs playing a role in verification of methods or data, to ensuring the availability of essential activity data.

3.3.1 Official statistics in the context of the greenhouse gas emission inventories

194. Official statistics play a key role in the process of monitoring social, economic and environmental development. NSO's role is to provide high-quality data necessary for management and decision-making. NSO data illustrate the effectiveness of the actions taken and progress in the implementation of strategic goals set by the government.

195. One of the areas of interest of official statistics are annual inventories of GHGs carried out by individual countries, which contribute to shaping climate policy and allow tracking progress in achieving national and international climate goals. The process of developing the GHG inventory and its reporting is defined by international agreements, which makes the data comparable on a global level.

What role can national statistical offices play in the inventory of GHG emissions?

196. The strengths of the official statistical system, in general, and the NSO in particular, have been discussed in Section 2.2.2.

197. First, official statistics in national statistical institutes follow international rules, adhere to high standards of professional independence and quality, and are committed to using robust, transparent and consistent methodologies. They are obliged to ensure the highest possible quality of produced statistical data. NSOs may have domestic standards for official statistics that can support the production of the inventory. For example, in New Zealand the Principles and Protocols for Producers of Tier 1 Statistics and release practice polices are used to ensure that the inventory is released using the principle of equal access and maintains independence. The Generic Statistical Business Process (GSBPM) model used by statistical offices may also be a guiding framework for inventory compiling agencies (UNECE, 2019a).

198. Second, NSOs are qualified to review and assess the usefulness of classification systems, registries, definitions and statistical frameworks. This can be particularly helpful in the early stages of setting up a GHG inventory when the IPCC guidance for identification of key categories and data collection methods is being implemented. NSOs could help assess the feasibility of meeting new inventory data needs and avoid creating overlapping data systems by evaluating available data.

199. Thirdly, NSOs are important and vital data providers in many very diverse areas. Most NSOs have a coordinating role in the national statistical system. They can therefore provide data on population, GDP, energy production and consumption, agriculture, forestry, mining, waste generation, manufacturing, transport, health and many more. NSO may make available such data, in an aggregated form, based on statistical surveys and from administrative sources, for the purposes of climate change, including GHG inventories.

200. Fourth, NSOs are experts in developing new data and ensuring the high quality of all the statistics they produce. They also play a key role in ensuring that these new data and related methodologies become part of the national statistical process, thus ensuring long-term stability. In the event of data gaps, NSO can support using its experience, knowledge and resources. This is of

particular importance in the context of the constantly evolving topic of statistics related to climate change.

201. Fifth, because of their responsibilities for macroeconomic indicators, NSOs can integrate different socio-economic data with information on emissions to produce emission estimates at the same time as publishing, for example, GDP indicators. Since climate change connects many areas, it requires a wide range of activity data across different statistical domains. As NSOs have information from all areas, their involvement in the work on emission inventories and related topics is necessary.

202. NSOs are themselves responsible for the production of essential data for the compilation of GHG inventories, e.g. on energy, industry, agriculture, forestry, transport, international trade, land use and land cover, sewage and waste. The inventories generally need data at more detailed level than are published or made publicly available by the NSOs as official statistics. For example, for the emission estimation for the energy and industrial processes sectors, operational unit/plant-specific data are needed, and it is not a question of simple multiplication of official statistics with emission factors. NSOs can use their experience in making more detailed data available in a secure manner and respecting confidentiality, and by doing so they can contribute to improving significantly the estimation quality.

203. One of the most important features of the inventory compilation methodology is the principle of recalculating data for the entire time series in case of obtaining new information or applying a new methodology. This means that the GHG inventories are subject to a systematic updating process, i.e., an annual revision of the entire time series. NSOs typically have clear revision policies, which are communicated to the public, and can share that expertise to support dealing with the regular revision of the inventories.

204. In some countries, NSOs are under-represented in national GHG inventory processes, and UNFCCC reviews and audits. The same may also apply to ministries of energy or energy agencies, which may lead to important data already collected being replaced by estimates or by less authoritative data. Most experts in the IPCC work on guidelines for developing national GHG emission inventories come from academia or national GHG agencies. This is a fundamental gap as the national inventory process relies heavily on data that is or can be produced sustainably by NSOs and other producers of official statistics. The role of NSOs as coordinators of statistical programmes could be leveraged to bring not only the NSOs but all relevant producers to the process. In addition, this process could benefit from NSO's methodological expertise and experience.

205. Close cooperation with NSOs would strengthen a sustainable national GHG inventory process and thus:

(a) The quality of GHG emission estimates would be improved;

(b) Data collection and their quality would be improved;

(c) Cooperation would be established between inventory compilers and statisticians in NSOs;

(d) The burden of data collection on data producers and respondents would be reduced;

(e) Consistency between different reports would be improved.

3.3.2 Tracking of nationally determined contributions

206. As described in section 3.2, each Party shall provide the information necessary to track progress in implementing and achieving its NDC, which will need to be included in the BTRs.

Opportunities to engage NSOs into NDC tracking

207. Involvement in tracking and reporting NDCs' progress would require increased human and financial capacities in both developed and developing country Parties. When there is a lack of dedicated institutions specialized in environmental data collection and climate policies management, NSOs could facilitate obligatory reporting on NDCs. Much of the information is already collected and

processed by NSOs and expanding their area of expertise would be more efficient than establishing new institutions. In other cases, NSOs could play a data stewardship role and help build strong co-operation with governmental agencies and bodies responsible for climate policy. Both approaches would need to extend the scope of activities and data collected by NSOs, along with the guidelines for NDC tracking and reporting.

3.3.3 Challenges facing statistical systems in the context of climate change statistics

208. Despite the many competencies that the statistical system brings to measuring climate change-related issues, existing official statistics are not always as useful for climate change analysis as they could be. Existing statistics have not been developed directly for the analysis of climate change. Therefore, the statistical system must continue to evolve to better respond to the need for statistics related to climate change. This also applies to data in the GHG emissions inventories.

209. In many countries, NSOs may not be responsible for much of the environmental statistics, even though they are mainly produced as part of the national statistical system. For example, environmental reports and biodiversity measurements are typically the responsibility of environmental ministries or specialized agencies. However, in most countries, the overall coordination of the national statistical system is entrusted to the NSO. One of the challenges for the development of climate change statistics is therefore to establish effective coordination and division of labour between different national authorities in the environmental area as well.

210. To acquire data according to the standard quality guidelines a lot of time and resources should be spent on setting up the entire statistical process, including data collection, methodology development, and capacity building. A timely and regular dialogue between the inventory community and NSOs, at national and international level, is needed to provide the right (activity) data. In general, it is underestimated how much time and resources an NSO has to invest to set up new or improved data flows.

211. Moreover, one of the challenges that NSOs faces in the context of emission inventories is the alignment of concept, methods and classifications with those used in official statistics in order to integrate and generate statistics in the most efficient and effective manner.

212. In addition, it will be a challenge to integrate and develop new statistics supporting the analysis of climate change in the statistical system, such as monitoring the shift towards greener consumption and production patterns.

3.3.4 Institutional arrangements

213. In the context of this chapter, a 'national system' can be defined as the institutional, legal and procedural arrangements within a country to estimate and report climate-change related information. This information includes GHG inventories, projections and policies and measures, to mention but a few. As such, institutional arrangements include the interactions between agencies within the country that are involved in the preparation and reporting of climate-change information, including, but not limited to, the agencies that are mandated by law to report on energy, industry, agriculture, forestry, transport, international trade, land use and land cover, sewage and waste.

214. There are several examples of how to improve collaboration and achieve better data coherence:

(a) A good practice is to have joint working groups between, for example, inventory compilers, and those responsible for energy statistics at the national level. This type of activity not only improves the quality of the energy data in the inventories and the energy data reported by the statistical offices, but also the coherence between both data sets, and improves the cooperation between the different groups. Such working groups may be helpful in all cases where there is more than one agency involved in data collection and/or compilation, including in the preparation, reporting and verification or review of the national inventory. Another specific area for collaboration to be explored is between the experts responsible for the LULUCF emissions and removals in the inventories and SEEA ecosystem accounts in the NSOs; which could collaborate to avoid double work, harmonize their methodologies and create synergies in eliminating data gaps;

(b) Legal provisions for specifying the responsibilities across the different agencies of the national system for different statistical domains can help in ensuring consistency between data submitted for the different national and international reporting mechanisms;

(c) The areas of collaboration could include the avenues for improvement for instance, in methods, (micro) data access and sharing;

(d) There is also broad agreement on the added value of international and regional cooperation between countries with similar circumstances or even the same language.

215. Governments are confronted with the challenge of designing coherent policies that can simultaneously accelerate economic growth, and, at the same time, help adapt to or mitigate climate change. To successfully achieve both objectives, countries need (i) sound institutional arrangements for policy integration; and (ii) analytical capacities to assess policy options (also quantitatively).

216. It is important to highlight that UNFCCC/IPCC guidance on national systems is not prescriptive. Countries are free to decide on the institutional setting that allows the country to report and deliver sustained quality improvements in the reporting while considering their national circumstances.

217. National systems are the bedrock of the whole MRV and transparency framework, which is why, in the BTR under the Paris Agreement countries are also required to report on the national circumstances and institutional arrangements. The MPGs of the ETF break down the reporting of this information into several chapters: GHG inventories, tracking of progress, climate impacts and adaptation, and finance. Although establishing robust institutional arrangements is essential and can be hard, the reporting requirements related to national circumstances and institutional arrangements are rather basic and can be found in decision 18/CMA.1 (MPGs).

218. The current section focuses on the key elements of robust institutional arrangements that allow the information to be reported and improved over time. These key elements are based on the experiences and perspective of a national inventory compiler, but also build from experiences shared by UNFCCC Parties at the Multilateral Assessment and Facilitative Sharing of Views workshops. Some of these elements, if not all, can be extrapolated and/or may be also applicable to other countries and areas of reporting relevant to the Enhanced Transparency Framework under the Paris Agreement.

219. At the outset, it is important to highlight that national inventory systems are not only relevant for international reporting purposes but also for national policy making and policy evaluation. This means that MRV and transparency systems should not only be designed to deliver reports to the UNFCCC but should be used for domestic policy making and other national priorities.

220. Figure 3.4 summarizes some of the key elements that have led and will continue to lead to GHG inventory quality improvements in the European Union, which may be applicable to many countries. Each of these elements is explained below.

Figure 3.4.
Basis for quality improvements in the greenhouse gas inventory of the European Union - key elements of a robust national system

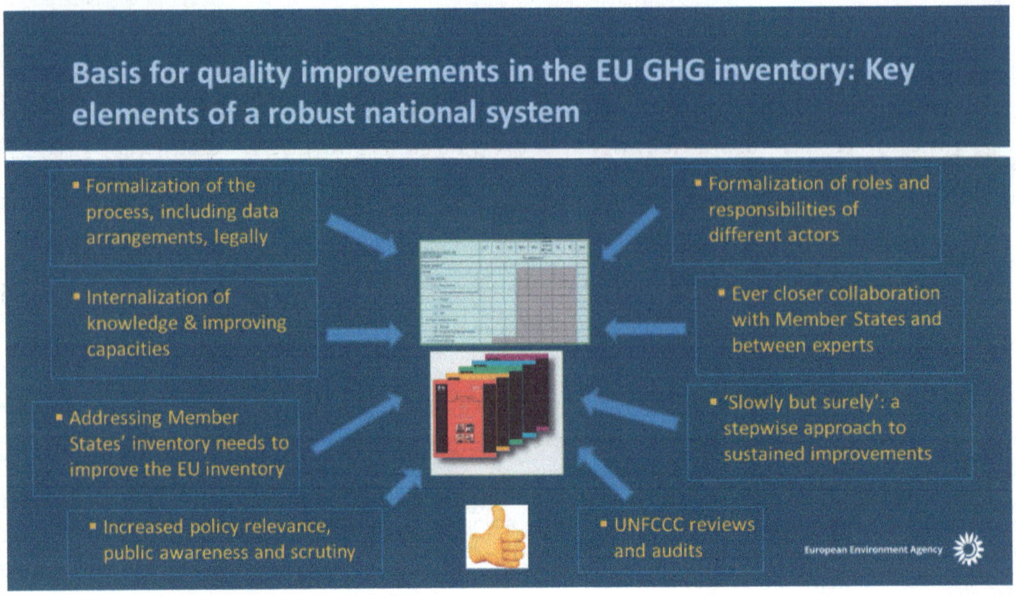

Source: European Environment Agency

Legal formalization of the process, including data arrangements

221. A key and essential element of a good MRV system is to ensure there are mechanisms at national level to ensure data sharing between the compilers and the agencies producing activity and other relevant datasets. One way is to develop legislation that allows the compilers to access all the data they need for the estimation, reporting and continuous improvement of inventories and other indicators, while respecting any confidentiality clauses.

Formalization of roles and responsibilities of different actors

222. Another essential element linked to ensuring effective data arrangements, is to formalize the role of the different actors involved in inventory preparation. The who does what, how and when. Formalization, including providing necessary budget, is essential to provide certainty about the roles and responsibilities of each institution during inventory compilation and review year after year.

Internalization of knowledge & improving capacities

223. Internalizing knowledge and improving capacities in the different activities, including data collection and the quality assurance/quality control of these data, are very important in a system that should deliver improvements in reporting over time. This is because if we do not internalize what is learnt, knowledge goes away. Retaining this knowledge would set the basis for future quality improvements.

Closer collaboration with Member States and between experts

224. Another key element of a national system is to seek closer cooperation and collaboration between countries and between experts. Improving collaboration and cooperation are also important for shared ownership and quality improvements. This can include a) improved quality of estimates by peer reviews or simple cross-checking of the reported information; and b) better consistency or data coherence between reporting obligations, including closer alignment with official statistics. The collaboration could be achieved through joint working groups, for example, between the reporting agencies and those responsible for official statistics.

Addressing lower-level data requirements to improve the national or regional inventory

225. In certain cases, the reporting agency may not produce its own activity data, and rather incorporate emission estimates or other statistics received from agencies at a lower regional level. Investing in improving local level data availability would benefit the national inventory. In a similar sense, improved Member State data availability would benefit the EU inventory.

Increased policy relevance, public awareness and scrutiny

226. The above-mentioned factors can contribute significantly to ensure steady quality improvements in the reporting of climate targets, such as the 2030/2050 national determined contributions, the Global Stocktake, and climate neutrality goals. But there is also an increasing number of public and journalist requests related to climate change, and its links to policies and measures. Because of these, it is important to engage different national stakeholders outside the government like the corporate sector, civil society and academia, throughout the reporting process, including as part of the MRV and transparency process.

UNFCCC reviews and audits

227. UNFCCC reviews and audits have also substantially improved the reporting processes. The implementation of the recommendations of these reviews, though demanding to start with, form the basis for increased trust between countries, credible climate policies, and the assessment of progress towards national, international and global climate targets and objectives.

'Slowly but surely': a stepwise approach to sustained improvements

228. The areas of improvement are not only identified by UNFCCC reviews and audits, but also by Parties themselves. To guarantee a stepwise sustainable implementation, 'slowly but surely', of the identified improvements also capacity-building needs are identified. Parties are expected to report in subsequent BTRs on how they are addressing or intend to address the identified areas of improvement and capacity-building. The reporting and review process under the Enhanced Transparency Framework (ETF) is aimed at facilitating continuous improvement over time.

3.4 Conclusions and recommendations

229. The Paris Agreement is built around national climate targets (nationally determined contributions), plans to achieve them, and tracking and reporting of countries' progress towards the implementation and achievement of these plans conducted under the enhanced transparency framework (ETF). The ETF will serve as the primary vehicle for the governments to demonstrate progress in implementation and achievement of their NDCs. It needs to be prepared and communicated every five years. Also every five years, governments will come together to participate in the global stocktake - a process of periodically taking stock of the implementation of the Paris Agreement to assess the collective progress towards achieving the long-term goals of the Agreement.

230. Meeting the reporting obligations under the Paris Agreement is a process, and the establishment of robust and sustainable institutional arrangements is a key enabling factor. This includes building capacity and expertise within a country so that necessary information can be collected and provided in a timely and reliable manner. The UNFCCC/IPCC guidance on national institutional arrangements is not prescriptive, and countries are free to decide on the institutional setting that allows a country to report and deliver sustained quality improvements in the reporting while considering their national circumstances.

231. Key elements of a robust national system include:

(a) Formalization of the process legally, including data arrangements;

(b) Formalization of roles and responsibilities of different actors;

(c) Internalization of knowledge and improving capacities;

(d) Ever closer collaboration between experts;

(e) Addressing lower-level data requirements to improve the national or regional inventory;

(f) Increased policy relevance, public awareness and scrutiny;

(g) UNFCCC reviews and audits;

(h) 'Slowly but surely': a stepwise approach to sustained improvements.

232. NSOs have a role to play in the ETF, even if the exact role and degree of involvement may differ by country. At minimum, NSOs are responsible for producing a significant share of the input data to GHG inventories, such as statistics on energy, agriculture, waste and industrial production. NSOs can also be formally involved in the reporting, e.g. in data quality assurance or by providing more detailed data to the compilers in a secure manner. In a few countries, NSOs are responsible for at least part of GHG emissions estimation or inventory compilation. In any case, NSOs should be actively involved to understand how data are used, discuss quality issues, and improve data collection if needed. Furthermore, NSOs produce a lot of statistics that can be used to inform or monitor actions needed to implement and track the NDCs, which countries submit under the Paris Agreement.

233. A good practice is to have joint working groups between, for example, inventory compilers, and those responsible for energy statistics at the national level. This type of activity not only improves the quality of the energy data in the inventories and the energy data reported by the statistical offices, but also the coherence between both data sets and cooperation between the different groups.

234. To build capacity related to the climate reporting process, with a primary focus on the data in the Common Reporting Tables (CRTs), NSO staff can also be trained as UNFCCC reviewers, who will be needed. This would allow them to learn about the specific data needs in the climate reporting process, which is not easy to understand without being practically involved in some way and establish relationships with the climate community. Involving NSOs in the UNFCCC review process facilitates addressing (micro)data gaps and improving the statistical quality of the data provided in the Enhanced Transparency Framework. Whether NSOs can allocate budget for the training of UNFCCC reviewers depends on national circumstances, such as the NSO mandate and other training priorities.

235. Finally, improving availability of internationally comparable data will facilitate the subsequent collective assessments of progress - the Global Stocktake.

Chapter 4 Statistics for climate change mitigation policymaking

4.1 Introduction

236. This chapter provides guidance on the role that NSOs can play in in informing and monitoring climate change mitigation policies, put in place nationally, or subnationally, to achieve the Paris Agreement mitigation goal. The scope covers not only measuring mitigation actions themselves, but a broad spectrum of data needs related to designing, implementing and monitoring climate change mitigation policies.

237. Section 4.2 introduces the official definition of climate change mitigation and presents examples of mitigation policies on global, regional and national levels and related data needs. Section 4.3 describes how NSOs can contribute by producing data complementing official GHG inventories (following both production and consumption approach), from collection of primary data through development of statistics and indicators to their dissemination. Examples focus on existing indicators, classifications, and statistical frameworks. This section also highlights the importance of energy statistics. However, the financial aspects of climate mitigation are discussed in Chapter 8.

238. The available statistics, i.e. the supply side (section 4.3; perspective of NSOs and other data producers), are confronted with policy data needs, i.e. the demand side (section 4.2; perspective of policy makers and data users) to reveal the data gaps. Based on the comparison, recommendations are made how to close the gaps and further strengthen the role of NSOs in the mitigation policy process.

4.2 Definition and policy context of climate change mitigation

4.2.1 Definition of mitigation

239. In its Sixth Assessment Report, the IPCC gives the following definitions (IPCC, 2022b):

> ***Mitigation (of climate change):*** *A human intervention to reduce emissions or enhance the sinks of greenhouse gases.*
>
> ***Mitigation measures:*** *In climate policy, mitigation measures are technologies, processes or practices that contribute to mitigation, for example, renewable energy technologies, waste minimisation processes, and public transport commuting practices.*

240. The definition of mitigation in the UNEP glossary (UNEP, 2022) is essentially the same, though additional examples of mitigation measures are given, e.g., using fossil fuels more efficiently for industrial processes or electricity generation, switching to solar energy or wind power, improving the insulation of buildings, and expanding forests and other 'sinks' to remove greater amounts of CO_2 from the atmosphere.

241. Such mitigation policies and their regulatory measures, such as low carbon energy technologies, energy efficiency measures, reform/reduction of fossil fuel subsidies, energy and transport taxes, emissions trading schemes, carbon capture and storage, can be measured and have quantifiable effects. Relevant statistics in this context include data on the GHG emissions, the underlying (sectoral) activities, but also on broader socio-economic phenomena such as population growth, urbanization, industrialization, etc. that also are emission drivers.

4.2.2 Overview of mitigation policies

242. Mitigation policies vary from country to country. A harmonized grouping can be used to facilitate an international overview. Example of such grouping is shown in Figure 4.1. The main categories are: (I) sectoral policies, (II) cross-sectoral policies, and (III) international policies and the

priority areas: (a) Energy supply, (b) Transport, (c) Buildings, (d) Industry, (e) Agriculture, (f) LULUCF, and (g) Waste, with a division into (1) market-based instruments and (2) non-market-based instruments.

Figure 4.1
Possible grouping of mitigation policies

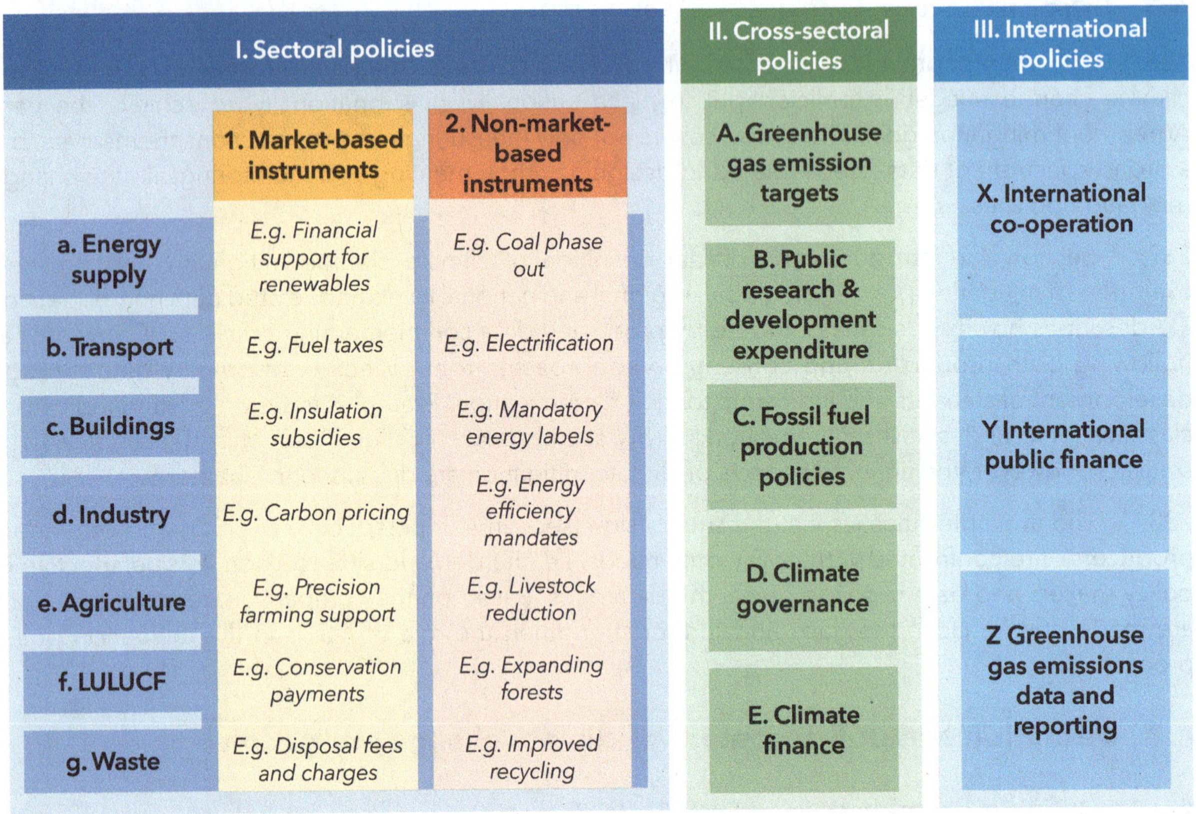

Source: Based on Nachtigall, D. et al. (2022); modified by the authors.

243. Market-based instruments are policy instruments that use markets, prices and/or other monetary means to provide incentives for producers and consumers to reduce or eliminate environmental and other externalities. Examples are carbon pricing like emissions trading schemes, carbon and fuel taxes, and governments' reform of fossil fuel support (in all sectors), financial support for renewable energy sources and tradeable renewable energy certificates (a. Energy supply), congestion charges for passenger cars in urban areas (b. Transport), financing mechanisms for energy efficiency (c. Buildings; d. Industry), fertilizer taxes (e. Agriculture), payments for conservation (f. LULUCF), and landfill taxes (g. Waste).

244. Non-market-based instruments are instruments that work through the imposition of certain obligations or by installing non-monetary incentives to change behavior. Examples are bans or phase outs of unabated existing coal power plants (a. Energy supply), fossil-based passenger cars (b. Transport), and fossil fuel heating systems (c. Buildings). Some energy-related examples are minimum energy performance standards for electric motors (d. Industry) and for appliances like lighting, refrigerator, freezer and air conditioner (c. Buildings), mandatory energy labels for such appliances (c. Buildings) and for light duty vehicles (b. Transport), energy efficiency mandates for industrial facilities (d. Industry) and building energy codes (c. Buildings). Other examples are planning for renewables and air pollution standards for unabated coal power plants (a. Energy supply), speed limits on motorways, emission limit values, and increased share of rail infrastructure (b. Transport), fertilizer and nitrate regulations (e. Agriculture), forest as protected area (f. LULUCF) and extended producer responsibility schemes (g. Waste).

245. For cross-sectoral policies, climate actions are linked to: (A) GHG emission targets (short term: nationally determined contributions; long term: net-zero targets), (B) Public research & development

expenditure (in low-carbon technologies), (C) Fossil fuel production policies (e.g., bans and phase outs of fossil fuel extraction, reform of fossil fuel producer support, and policies to reduce fugitive methane emissions), (D) Climate governance (e.g., an independent climate advisory body), and (E) Climate finance (e.g., mandatory disclosure of climate risk information for companies).

246. For international policies, climate actions can be grouped into: (X) International co-operation (e.g., participation in international climate agreements and climate initiatives, under UNFCCC umbrella, and participation in international emissions pricing from aviation and shipping), (Y) International public finance (e.g., banning government's export credits for new unabated coal power plants and banning public finance for fossil fuel infrastructure abroad), and (Z) GHG emissions data and reporting.

247. It should be emphasized that the overview of policy measures, in Figure 4.1, is not an official classification and differs from the one used under UNFCCC and Paris Agreement for the GHG inventories and projections as well as reporting on policies and measures under these agreements. It serves in this chapter as an example of a grouping for reference in the next section. A comprehensive stocktaking and mapping of climate change mitigation policy instruments to their emissions base is being carried out by the OECD Inclusive Forum for Carbon Mitigation Approaches (IFCMA) in 2024 and beyond.[21] These IFCMA activities will enhance the understanding of the impact of the full spectrum of national and subnational carbon mitigation approaches deployed around the world to reach the net-zero goals. IFMA also supports the development of more accurate, timely and granular carbon intensity metrics, which can play a crucial role in emission reduction efforts by governments, businesses and consumers.

4.2.3 Mitigation policies in nationally determined contributions

248. The Paris Agreement requests each country to outline and communicate their post-2020 climate actions, known as their nationally determined contributions (NDC). These NDCs not only contain information on targets, policies and measures for reducing national emissions, but also information on either the need for, or the provision of, finance, technologies and capacity building for these actions.

249. The NDC Synthesis Report summarizes the NDCs as recorded on 23 September 2022 (UNFCCC, 2022a). Most Parties (90 per cent) provided quantified mitigation targets, expressed as clear numerical targets, while the rest (10 per cent) included strategies, policies, plans and actions for which there is no quantifiable information as components of their NDCs.

250. Almost all Parties (99 per cent) explained, using different metrics, how they consider their NDCs to be fair and ambitious in the light of their national circumstances. Metrics include, e.g., share in global emissions, level of per capita emissions, vulnerability to adverse impacts of climate change, historic and current responsibility, climate justice, capabilities, mitigation potential, and cost of mitigation actions.

251. Most Parties (87 per cent) indicated that domestic institutional arrangements are key element of coordination, planning and implementing climate change policy and action at (inter)national level and fostering public participation. Most Parties (79 per cent) referred to formal arrangements for domestic stakeholder consultations and engagement processes, which were generally conducted in an inclusive and participatory manner.

[21] https://www.oecd.org/en/about/programmes/inclusive-forum-on-carbon-mitigation-approaches.html

Figure 4.2
Specific priority areas and frequently indicated mitigation options in nationally determined contributions (share of Parties; 2022)

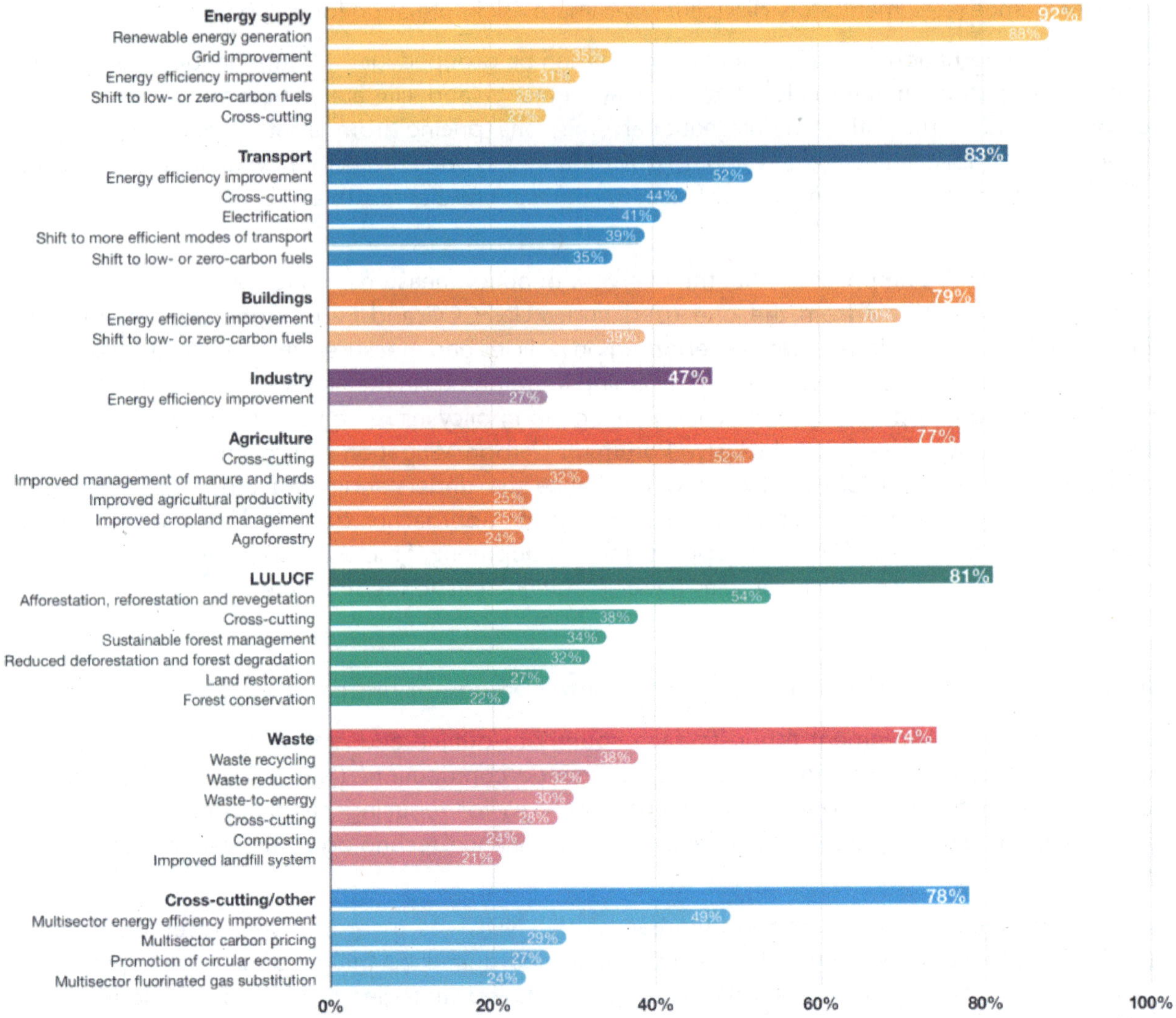

Source: UNFCCC (2022a).

252. Almost all Parties (92 per cent) outlined in their NDCs domestic mitigation measures in the priority area of Energy supply (see Figure 4.2). The percentages for the other priority areas are Transport (83 per cent), LULUCF (81 per cent), Buildings (79 per cent), Cross-cutting/other (78 per cent), Agriculture (77 per cent), Waste (74 per cent), and Industry (47 per cent). The highest sub-area percentages belong to Renewable energy generation (88 per cent) and Energy efficiency improvement (70 per cent for Buildings; also quite high percentages for Transport, Cross-cutting/other, Energy supply, and Industry). Activities to increase the renewable energy generation are combined with mitigation options to phase down unabated coal power generation, to phase out inefficient fossil fuel subsidies, and to reform fossil fuel subsidies.

253. Quantitative mitigation targets were provided most frequently for Energy supply (57 per cent), LULUCF (43 per cent), and Cross-cutting/other (42 per cent), while they were much less frequently indicated in the other priority areas (5-19 per cent). The highest sub-area percentage for quantitative mitigation targets belongs to Renewable energy generation (54 per cent).

Figure 4.3
Mitigation options in NDCs (share of Parties; 2022); in parentheses behind each option: net emission reduction potential (Gt CO_2-eq/yr)

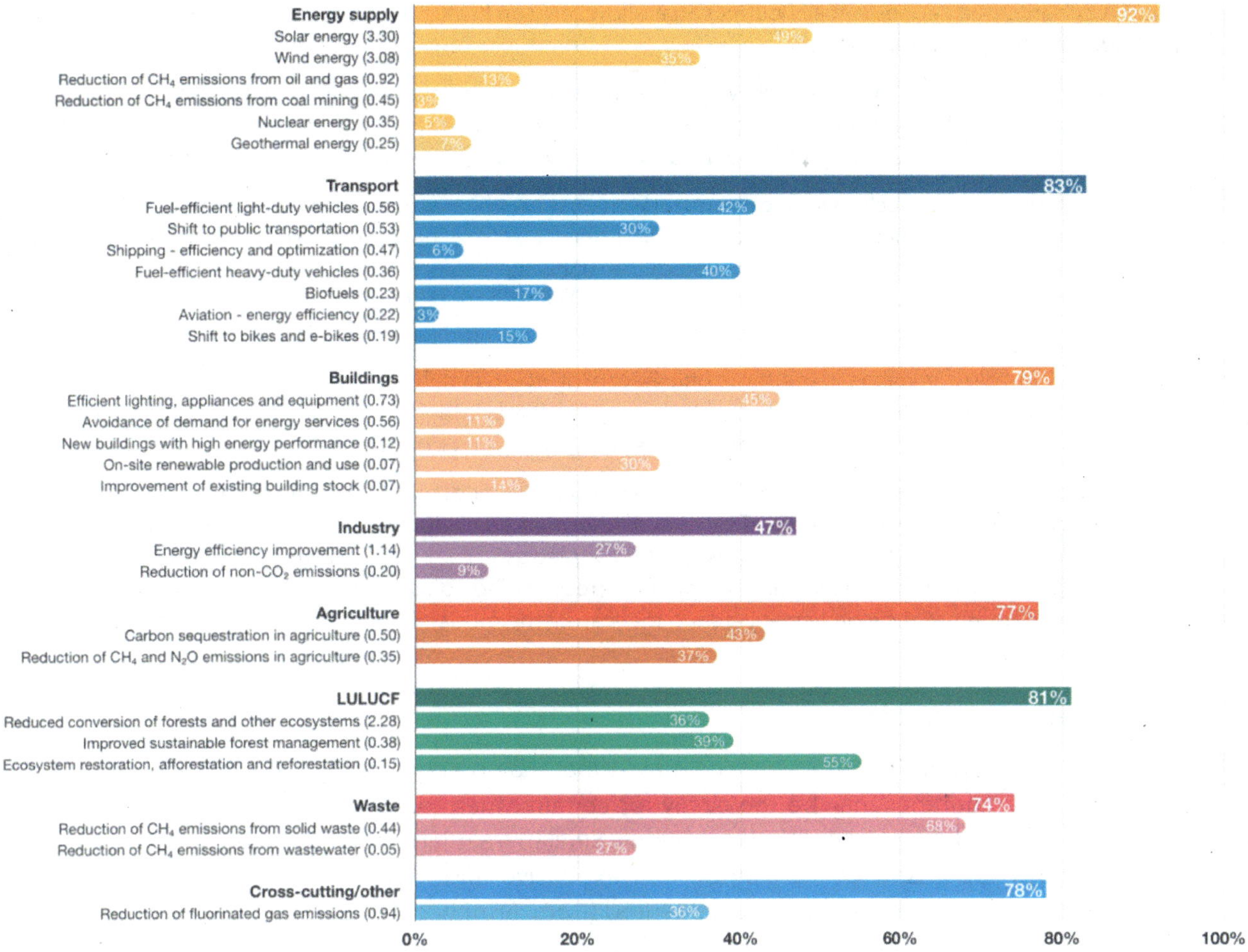

Source: UNFCCC (2022a).

254. Cost-effective mitigation options, i.e., costing less than '\$20 per ton CO_2-eq', with the highest estimated net emission reduction potential include (Figure 4.3): Solar energy (3.30 Gt CO_2-eq/yr), Wind energy (3.08 Gt CO_2-eq/yr), and Reduced conversion of forests and other ecosystems (2.28 Gt CO_2-eq/yr). This corresponds to potential emission reductions of 6, 5, and 4 per cent, respectively, with respect to the total global GHG emissions of 2019. Other cost-effective mitigation options represent estimated potential reductions of 2 per cent or lower. The mitigation option reduction of CH_4 emissions from solid waste (0.44 Gt CO_2-eq/yr) is outlined in NDCs by the highest percentage of Parties (68 per cent), though it represents an estimated potential reduction of less than 1 per cent.

255. Most Parties (74 per cent) identified capacity building as a prerequisite for implementation of their NDCs. Capacity-building needs were identified for formulating policy, integrating mitigation into sectoral planning processes, providing NDC related information, and accessing finance. Almost all Parties (92 per cent) provided information on finance as a means of NDC implementation, with most (69 per cent) characterizing finance in terms of international support needed and some (22 per cent) mentioning finance in relation to domestic implementation only.

256. In parallel to official emission trading systems, such as the EU Emissions Trading System (ETS)[22], countries can set up voluntary carbon markets, in which NSOs can play a role. See Box 4.1 for more information.

[22] https://climate.ec.europa.eu/eu-action/eu-emissions-trading-system-eu-ets_en

Box 4.1 Carbon markets

A carbon market is a specialized type of financial market, through which carbon credits can be bought and sold. Carbon credits are essentially permits that allow the purchaser to emit a certain amount of carbon dioxide or other greenhouse gas.

Article 6(2) of the Paris Agreement allows Parties to enter into mutual agreements to trade mitigation outcomes that can be reported and accounted for under an international framework. In other words, mitigation efforts in one country could be used to fulfil NDC targets in another country.

Article 6(4) provides a structure for a carbon credit market on which greenhouse gas emission reductions or removals may be transferred internationally. This puts pressure on national governments to design their own national carbon market frameworks. One of the main challenges in such voluntary carbon markets is the lack of standardization, integrity and transparency. Without clear standards for carbon credits, it can be difficult for companies to know whether they are truly reducing their emissions.

One possible role for national statistical offices can be to maintain and update the national carbon registries. Carbon registries are a vital part of ensuring the integrity and credibility of the voluntary carbon market. A carbon registry is a system used to generate and track carbon credits throughout their lifecycle.

Article 6(2) sets out the requirement for Parties to apply corresponding adjustments to carbon credits that are used for trading between Parties. This is to ensure that an emission reduction is not counted as a mitigation by two different Parties towards their Paris pledges. To maintain such live registry and keep them updated to avoid double counting under the corresponding adjustments, the role of national carbon registries can be very critical particularly during the start of the model.

4.2.4 Country examples

257. The mitigation policies vary by country. What may be considered a cost-effective mitigation option for most countries may be too costly for certain other countries, depending on local, socio-economic, technological and political conditions. Similarly, a mitigation option that most countries do not consider to be cost-effective may be less expensive and therefore feasible for some specific countries. Some country examples are presented (from Armenia, Canada, the Netherlands, Pakistan and Spain; see 4.2.5) to show similarities and differences in national mitigation policies. It is a snapshot and certainly not complete.

258. In Energy supply sector national mitigation policies generally focus on measures like coal phase out and increasing use of renewable energy sources (solar, wind, hydro and geothermal). Many renewable energy sources and electrified technologies are now economically on par with the fossil fuel-based equivalents. Related mitigation measures include improvement of electricity infrastructure, energy storage alternatives, hydrogen strategies, and programmes/funding for clean fuels and clean energy technologies. Local conditions can be taken into account using Regional Energy Strategies, e.g., by supporting citizen participation in the form of energy communities to increase public acceptance of renewable energy projects.

259. In Buildings sector the focus is on energy efficiency improvements. Elements of mitigation measures are Minimum Energy Performance Standards, funding for homeowners to improve existing building stock, near zero emission building codes for new houses, on-site renewable production and use, and less use of fossil fuels for heating purposes (more electric heat pumps).

260. In Industry the focus is not only on technical energy efficiency improvements (energy savings, more renewables, and cleaner technologies and processes), but also on the reduction of CO_2 emissions by carbon pricing and by innovative projects related to carbon capture and storage (CCS). National governments facilitate the largest industrial emitters through investments in their emission reduction projects towards net zero by 2050.

261. In Transport the focus is on the electrification of road traffic. This requires investments in an infrastructure of charging stations. All kinds of zero-emission vehicles incentives (mostly financial) are implemented, combined with a phase out of fossil-based transport. In order to have less road traffic

public transport can be optimized and employees can be encouraged to work more often at home. There is also a need to shift to more efficient and sustainable modes of transport and to replace fossil fuels by bio-alternatives.

262. In agriculture the main focus is on reducing the CH_4 and N_2O emissions from livestock farming and fertilizer application. A more sustainable operational management can be achieved by innovation subsidies and on-farm environmental stewardship programmes. Energy savings, e.g., in greenhouse horticulture, will reduce the agricultural CO_2 emissions. Other measures include increased use of renewable energy sources and investments in clean technology and geothermal energy projects.

263. In LULUCF the focus is on improved sustainable forest management. Measures can consist of, e.g., longer rotations, less intensive harvests, continuous-cover forestry, mixed stands, more adapted species, selected provenances, and high-quality wood assortments (see paragraph 7.4.2.3 in IPCC, 2022b). Current national policies include investments in planting additional trees, carbon sequestration, and wildfire prevention. Not all LULUCF mitigation measures are forest-related, e.g., greenhouse emissions from peat meadow areas can be reduced by raising the groundwater levels.

264. Cross-cutting mitigation measures are focused on finance and investment (e.g., carbon pricing, environmental taxes and green bonds) and the labour market (energy transition related jobs and skills). The financial sector can be a driving force behind national and international mitigation policies. It may help investors make more informed decisions in the mitigation domain. The financial sector can also accelerate the portfolio decarbonization and set guidelines for green banking. Other cross-cutting mitigation measures are the promotion of circular economy, the Global Methane Pledge, investments in Nature Based Solutions, and registries for, e.g., carbon footprints and sink projects. Finally, some countries need financial support from other countries to reach their full emission targets.

265. The Policies and Measures Database[23] of the International Energy Agency (IEA) includes access to information on existing or planned government policies and measures to reduce GHG emissions, improve energy efficiency and support the development and deployment of clean energy technologies. The policy information in the IEA database has been collected since 1999 and governments have an opportunity to review the policy information periodically. In the next section IEA database examples are included of existing national mitigation policies from the United States and China. However, at subnational level mitigation policies may differ, e.g., U.S. State Renewables Portfolio.[24]

266. The national mitigation policies from the United States and China are similar to the ones from countries like Armenia, Canada, Pakistan, the Netherlands and Spain. For the Energy supply sector it covers measures to further increase renewable energy share, to gradually improve grid resilience, and to accelerate energy storage innovation. For both Buildings sector and Industry the focus is on improvement of energy efficiency. It includes the Made in China 2025 strategy (green and smart manufacturing). Additionally, the United States provides tax credits not only for carbon storage, but also for enhanced oil recovery or other industrial uses of CO_2, provided emission reductions can be clearly demonstrated. In Transport the focus is on the electrification of road traffic (New Energy Vehicles; improved network of charging stations). The United States also invests in the development of new aircraft technologies and mentions the importance of key battery minerals in electric cars, along with the rare earth minerals used in wind turbines, and other materials used in clean energy. Finally, carbon pricing and investments in sink projects (carbon storage) are mentioned.

267. In general, all countries implement similar national mitigation policies, which vary due to specific national circumstances. Many of the mitigation measures are energy-related, except in LULUCF sector (e.g., sustainable forest management), in Agriculture (e.g., livestock farming and fertilizer application; non-CO_2) and for the non-CO_2 cases in the other sectors.

[23] https://www.iea.org/policies/about

[24] https://emp.lbl.gov/projects/renewables-portfolio

4.2.5 National mitigation policies by sector

268. Sections below present examples of national mitigation policies by sector. The following sectors are distinguished: Energy supply, Buildings, Industry, Transport, Agriculture, LULUCF, and Cross-cutting. Mitigation policies from Armenia, Canada, the Netherlands, Pakistan and Spain were collected by mail through the Task Force members in July-August 2022. Mitigation policies from the United States and China were extracted from the IEA Policies and Measures Database in June 2023. This overview is not complete – it is a snapshot and could change in the meantime.

4.2.5.1 Energy supply

Armenia

- Implemented, ongoing and planned mitigation actions: construction on solar power plants, upgrade of the hydro power plant, commissioning of small hydro power plants, and upgrade of distribution networks implemented by the electric networks.
- In accordance with the 2020 Energy Sector Development Strategic Programme to 2040, the Government intends to increase the share of solar energy generation in total to at least 15 per cent or 1.8 billion kWh by 2030. To this end, solar power plants with the capacity of about 1,000 MW, including autonomous ones, are planned to be constructed.

Canada

- Coal phase-out: Coal-fired power is currently the biggest source of emissions in the electricity sector. The Government of Canada has passed regulations to accelerate the phase out of unabated coal-fired electricity by 2030.
- Hydrogen Strategy: The Strategy lays out a framework for actions that will reinforce hydrogen as a tool to achieve our goal of net-zero emissions by 2050 and position Canada as a global, industrial leader in clean fuels.
- Energy Innovation Program: The program funds research, development, and demonstration projects, and other related scientific activities that advance clean energy technologies.
- Clean Fuels Fund: The fund is designed to de-risk the capital investment for building new or retrofitting or expanding existing clean fuel production facilities.

China

Guidelines on renewable energy power purchase management have been established. According to the national determined benchmark prices and the number of affordable acquisition hours, grid enterprises need to buy all renewable energy electricity within the full acquisition planning under the premise to ensure the safety of the power supply.

To ease government subsidies to wind and solar sectors, the National Development and Reform Commission launched in 2017 a voluntary trading mechanism for green certificates. The solar and onshore wind power producers were issued 1 MWh certificates to sell to private and state-owned businesses, who were encouraged to purchase them.

In order to implement decisions on the promotion of renewable energy development and energy revolution, the Ministry of Water Resources adopted 'Guidelines on promoting development of small hydropower projects'. The policy document outlines a plan for the development of small hydropower installations until 2030 with regard to environmental protection, based on technology improvements and best practices of plant construction, operation and management.

The Netherlands

- Coal phase-out. A ban to use coal in electricity production in 2030 and beyond.
- Increased use of renewable energy sources. In 2030 it is expected that 75-90 per cent of the electricity production will be based on wind energy (offshore and onshore) and solar energy (onshore).
- Development of Regional Energy Strategies to implement onshore renewable energy sources, i.e., this implies careful spatial integration with the support of stakeholders (including the people who live there).
- Improvement of the electricity infrastructure, which needs to be adapted to the increased use of renewable energy sources, to ensure the supply of electricity when it is needed. The flexibility of the

entire energy system can be improved by converting the excess electricity from renewable energy sources into hydrogen. More biogases might be produced.

Pakistan

- Pakistan has shown a desire to shift to renewables and gave a target of 60% energy transition to renewable resources by 2030.
- Moratorium on new coal power plants. No Power generation through imported coal by 2030.

Spain

- Coal and nuclear phase-down. The owner companies and public administration have agreed on a coordinated closure of the different installations according to their useful life.
- Increased use of renewable energy sources. In 2030 it is expected that 74 % of the electricity production will be based on renewable sources (and 100% in 2050). More than additional renewable 58 GW in 2030 and 179 GW in 2050 with respect to 2016. This deployment is boosted through auctions of renewable electricity production and other programs.
- Strong impulse to energy storage alternatives and to the deployment of renewable Hydrogen pilot and commercial solutions and other renewable gases.
- Promotion of energy communities and self-consumption. Improvement of the electricity infrastructure and the interconnections deployment with Europe.
- Measures to reduce administrative burden in the authorization procedures for renewable power plants projects, and specifically in relation to the connection to the grid.

United States

The Department of Energy has announced the availability of USD 5 billion and USD 3.25 billion in loans, to improve resilience and expand transmission capacity across the grid, as it prepares for a higher share of variable electricity generation from renewables.

The Department has announced the beginning of the design and construction of the Grid Storage Launchpad, a USD 75 million facility located at Pacific Northwest National Laboratory in Washington, that will accelerate energy storage innovation, boost clean energy adaptation and grid resilience.

The Department of Energy has announced up to USD 12 million for technologies that can make geothermal systems more efficient for clean, renewable energy production. The funding will target enhanced geothermal systems to unlock the potential for increased electricity generation using geothermal energy.

Funded under the Solar America Initiative, the Department has announced funding of up to USD 13.7 million, over 3 years, for university-led projects that will focus on developing advanced solar power technology manufacturing processes and products. The project aims for selected universities to leverage fundamental understanding of materials and solar devices to help industry partners advance manufacturing processes and products.

4.2.5.2 Buildings

Armenia

- Implemented, ongoing and planned mitigation actions: scaling-up investment in energy efficient building retrofits, energy efficiency in public buildings, household energy efficiency mortgage loans, investments for energy efficiency repair works and lighting in municipal buildings, and projects on lightening.
- The Government plans to gradually expand the implementation of the programmes that will contribute to more buildings having their own individual heating and hot water production systems based on renewable energy resources.

Canada

- Established the Greener Homes Grant which helps homeowners make their homes more energy efficient and grows domestic green supply chains.
- Launched the $1.5 billion Green and Inclusive Community Buildings program which provides funding to support projects that improve energy efficiency through retrofits, repairs or upgrades and new builds.

China

The key tasks of the Green and High-Efficiency Cooling Action Plan (2019) include: raising space cooling minimum energy performance standards at or above developed standards levels, support the development of more efficient cooling technologies and enhance cooling-related energy-saving retrofits.

The standard for energy consumption of buildings is issued by the Ministry of Housing and Urban-Rural Development in 2016. It is the first national standard of building operation energy consumption indicators, which sets the target of building energy-efficient operation. The standard provides constraint value and leading of building energy consumption indicators, covering residential buildings, commercial buildings, and north China heating.

The Netherlands

- Increased energy saving (insulation measures and improved energy labels) and less use of natural gas for heating purposes (more electric heat pumps and an increased use of heat networks). The local policies, coordinated by municipalities, are combined with national policy measures focused on the individual owners of houses and other buildings.

Pakistan

- Revising building codes.
- Minimum Energy Performance Standards for appliances like air conditioners and LED lights.

Spain

- Strong rehabilitation subsidies program for existing buildings and substitution of natural gas with low carbon alternatives like heat pumps. Near zero emissions building code for new houses.
- Promotion of energy self- consumption in buildings based on renewable energy.

United States

Better Buildings Accelerators are designed to demonstrate specific innovative policies and approaches, which will accelerate investment in energy efficiency upon successful demonstration. Each accelerator is a targeted, short term, partner-focused activity designed to address persistent barriers that stand in the way of greater efficiency.

The Programme on Assisted Housing Stability and Energy and Green Retrofit Investments provides grants or loans through the US Department of Housing and Urban Development's Office of Affordable Housing Preservation for energy retrofits and green investments to property owners receiving project-based assistance. Incentives may be provided for owners to undertake energy or green retrofits as a part of such grant or loan terms.

As part of Infrastructure Investment and Jobs Act, measures related to Energy Efficiency represent a total budget of $1.025 billion. This includes: (1) Energy Efficiency Revolving Loan Fund Capitalization Grant program to conduct commercial or residential energy audits or upgrades and retrofits (USD 250 million), and (2) Competitive grant program within the Building Technologies Office to enable sustained, cost-effective implementation of updated building energy codes (USD 225 million).

4.2.5.3 Industry

Canada

- Cutting Corporate Taxes for Makers of Zero Emissions Technology: The Government of Canada proposed in Budget 2021 to reduce by half the general corporate and small business income tax rates for businesses that manufacture zero-emission technologies.
- Clean Growth Program: In 2017, the Government of Canada launched a program to invest in clean technology research, development, and demonstration in the Canadian energy, mining, and forestry sectors. Program ended in March 2022.
- Strategic Innovation Fund: The Government of Canada has launched a Net Zero Accelerator to support the decarbonization and sustainable growth of Canada's largest industrial emitters through investments in the adoption of clean technology and processes that will dramatically reduce the GHG footprint of these industries by 2030 and create pathways to net zero by 2050.

China

The 'Key Work Plan for Industrial Energy Conservation Supervision in 2020' was formulated in order to implement the 'Energy Conservation Law' and 'Industrial Energy Conservation Management Measures', give full play to the supervision and protection role of energy conservation supervision, continuously improve the industrial energy efficiency and green development level, and promote the high-quality development of the industrial economy in line with the 'Industrial Green Development Plan'.

The 'Fund management guidelines for industrial transformation and upgrade' were set out to support the implementation of the Made in China 2025 strategy. The fund will support projects on, e.g., smart manufacturing, smart consumer goods and green manufacturing through direct grants, concessional plans, credit guarantees and insurance.

The Strategic Plan for Industrial Efficiency targets the metallurgical and (petro-)chemical industry. It involves equipment renovation, the design and implementation of process optimization, and management measures. China aims to improve energy efficiency and industrial competitiveness to the highest level or close to the worlds front-runners.

The Netherlands

- Implementation of a national CO_2 tax in 2021 and beyond to reduce the industrial emissions. This increases the original CO_2 price as set in the European Emission Trading System.
- Reduction of current industrial emissions by one third in 2030. To this end, the Dutch government facilitates the largest industrial emitters in the Netherlands in their reduction projects, such as CCS, by a timely availability of infrastructure, (innovation) subsidies, and permits.

Pakistan

- Minimum Energy Performance Standards for electric motors.
- Energy efficient practices in industries.
- Pakistan's National Action Plan on Sustainable Development Goal 12.

Spain

- Implementation of energy efficiency programs. Implementation of renewable energy alternatives and the use of renewable Hydrogen for process emissions reduction.

United States

The Industrial Facilities Initiative helps Federal agencies improve facility energy efficiency. In addition to improving energy efficiency, the Initiative helps Federal industrial facilities meet regulatory requirements. This includes Executive Order 13423, which requires Federal facilities, laboratories, and industrial buildings to reduce energy consumption 30% by 2015 compared to a 2003 baseline.

The so-called policy 'Section 45Q Credit for Carbon Oxide Sequestration' provides a tax credit for CO_2 storage. This policy is intended to incentivize deployment of carbon capture, utilization and storage, and a variety of project types that generate CO_2 emissions reductions are eligible. The 2018 changes to 45Q will provide up to USD 50 per tonne of CO_2 permanently stored and USD 35 per tonne of CO_2 used for enhanced oil recovery or other industrial uses of CO_2, provided emissions reductions can be clearly demonstrated.

4.2.5.4 Transport

Armenia

- Implemented, ongoing and planned mitigation actions: optimization of public transport, re-equipment of public electric transport, conversion of Yerevan public and private vehicles to compressed natural gas, improvement of road infrastructure, optimization of municipal transport and improvement of management efficiency and promoting fuel switching to electricity in transport.
- By the 2022 Decision No. 39 of the Eurasian Economic Commission, a quota was set for the import of vehicles with electric motors without the payment of import duties to the territory of Armenia, with the following quantities: 7,000 units for 2022, 8,000 units for 2023. The mentioned tariff privilege was envisaged until 1 January 2026.

Canada

- Public and Active Transportation: Government of Canada has made significant public transit investments to support sustainable modes of transportation.
- Zero-Emissions Vehicles Incentives: Rebates are made available to Canadians to offset the purchase price of zero-emission vehicles. Similarly, tax write offs are available for businesses making the transition with their fleets.
- Electric Vehicle Charging and Alternative Refueling Infrastructure: Investments are supporting the deployment of public charging and refueling stations across Canada, the demonstrations of next-generation and innovative charging and hydrogen refueling technologies, and the development of codes and standards.

China

The 2023 Subsidies on New Energy Vehicles (NEVs) are provided by the central government, and then allocated to the provincial-level departments of finance. Companies manufacturing NEVs, and meeting certain standards, can apply for the subsidy. The qualifying standards include economic, environmental and social performance of the company as well as customer satisfaction level.

The New Energy Automobile Industry Development Plan (2021-2035) focuses on strategies and targets to promote NEVs (including electric vehicles and hydrogen fuel cell vehicles). One of the main targets is to reach a fuel economy of 12kWh/100km for electric vehicles by 2025, and for NEVS to account for 20% of the new vehicle sales. In addition, the battery swapping system will also be improved.

The Netherlands

- Further electrification: By 2030, 100% of the new passenger car sales (and the entire bus fleet) should be zero-emission vehicles. Meanwhile, the infrastructure of charging stations is gradually being improved and the fossil fuels are replaced as much as possible by bio-alternatives.
- Reduced road traffic by making more use of public transport and bicycles and also by encouraging the employees to work more often at home.

Pakistan

- Bus Rapid Transit system in five big cities.
- Electric Vehicle Policy which aims to shift 30% on electric vehicles by 2030.

Spain

- Target for 5 million electric vehicles in 2030, subsidies and reduced taxes to achieve this target.
- No new internal combustion engine cars after 2040 included in the Law 7/2021
- Promotion of electrical charging points deployment
- Obligation on "Zero Emissions Areas" in cities with more than 50.000 inhabitants.

United States

The U.S. Departments of Transportation and Energy announced nearly USD 5 billion over 5 years that will be made available to make electric vehicle (EV) charging accessible to all Americans. The National EV Infrastructure Formula Program will help states create a network of EV charging stations along designated Alternative Fuel Corridors, particularly along with Interstate Highway System. A plan to review the state of domestic supplies chains, including batteries, key battery minerals and semiconductors used in electric cars, along with the rare earth minerals used in wind turbines, and other materials used in clean energy.

The U.S. is setting a goal of achieving net-zero GHG emissions from the aviation sector by 2050 in its publication of the United States Aviation Action Plan. To reach the goal, new aircraft technologies, sustainable fuel, and enhanced operations efficiency will be needed along with cutting airport emissions. The actions envisaged in the Plan include development of efficient aircraft and engine technologies, production and increased use of Sustainable Aviation Fuels and electrification for short-haul aviation.

The Renewable Fuel Standard Program established the first renewable fuel volume mandate in the U.S. The Program, set new reporting, registration, and compliance requirements for major refiners, fuel blenders, and fuel importers. The program was introduced with the aim of increasing the volume of renewable fuel required to be blended into gasoline.

4.2.5.5 Agriculture

Canada

- Launched the five-year, Canadian Agricultural Partnership, cost-shared with provinces and territories, that supports on-farm environmental stewardship programmes.
- Launched the Agriculture Clean Technology Program to support development and adoption of clean technology.

The Netherlands

- Providing innovation subsidies to achieve a more sustainable operational management in livestock farming. Examples: (1) studies on the effects of grazing and roughage rations on methane emissions in dairy farming and (2) a pilot study on manure processing in which the produced mineral concentrates were used as fertilizer substitute.
- Providing innovation subsidies to achieve energy savings in the greenhouse horticulture, including the implementation of several geothermal energy projects and increased use of renewable energy sources.

Spain

- Several measures (regulatory, incentives, innovation and subsidies) to adjust and reduce the application of nitrogen to the crop needs and for the management of manure and slurry from cows and pigs.

4.2.5.6 Land Use, Land Use Change and Forestry (LULUCF)

Armenia

- Afforestation and reforestation works (planting) where about 560 ha forest have been planted and restored. Since 2016, the Mainstreaming Sustainable Land and Forest Management in Mountain Landscapes of North-Eastern Armenia UNDP-GEF Project (2016-2020) has been implemented involving development of management plans of three forestry branches of Hayantar Expertise Centre of the Republic of Armenia in Tavush and Lori regions. The Addressing Climate Change Impact in Armenia through Enhanced Capacity for Wildfire Management UNDP-Russian Trust Fund Project's (2017-2020) aimed at addressing the urgent problem of mitigating forest and wildfire risks in a changing climate and improving and consolidating early warning.

Canada

- Established the Natural Climate Solutions Fund, which supports Canada's commitment to plant 2 billion trees, funds projects that conserve, protect and restore habitats that store and capture carbon and help farmers tackle climate change on agricultural lands.

The Netherlands

- Reduced GHG emissions in peat meadow areas by raising groundwater levels. To increase carbon sequestration the efforts are primarily aimed at a climate-smart forest management, but also other measures appear to be effective (like increasing the permanent grassland area). To achieve both emission reductions and carbon sequestration the Dutch government is providing (innovation) subsidies.

Pakistan

- Reforestation and reducing CO2 emissions from forest degradation, and Farm forestry as a carbon sink.
- Ten Billion Tree Tsunami Project will sequester 148.7 MtCO2e in next ten years.

Spain

- Regeneration of silvo-pastoral systems; Promoting poplars as replacements for agricultural crops in flood-prone areas; Creation of forest areas; Forestry activities to prevent forest fires; Promoting sustainable coniferous forest management, application of thinning schemes to increase carbon removals; Hydrological-forest restoration in areas at high risk of erosion; Promoting conservation agriculture (direct sowing); Maintenance of plant cover and the incorporation of pruning waste into soil for woody crops.

4.2.5.7 Cross-cutting - finance and investment

Canada

- Carbon pricing: Since 2019, Canada has a two-way carbon pollution pricing system in place.
- Green bonds: In 2022, the Government issued its inaugural green bond. These bonds will help finance green infrastructure and other projects that will help fight climate change and protect the environment.
- Climate-related financial disclosures: The government will work toward implementing climate-related disclosure to help investors make more informed decisions.

China

The Interim rules for carbon emissions trading clarify various definitions of the national carbon market, and cover the inclusion criteria of key emission entities, the setting and allocation of total allowances, trading entities, verification methods, reporting and information disclosure, supervision and penalties for breach. The rules enter into force on 1 February 2021. China's emission trading scheme does not have a cap. Emission allowances (permits) are allocated based on production. The Emission Trading System covers entities that emit over 26 000 tCO_2e per year.

Pakistan

- Water and Power Development Authority has launched 10-year Green Bonds and has raised $500 million for a hydro-energy project.
- Securities and Exchange Commission of Pakistan has approved national guidelines for green bonds that will encourage innovative financing mechanisms in several sectors for both adaptation and mitigation.
- The State Bank of Pakistan in 2017 approved green banking guidelines to promote environmental risk management within the commercial banks and encourage climate finance to reduce vulnerabilities.
- Pakistan has committed to reduce 15% of projected GHG emission through its own resource and requested the remaining 35% with the cooperation of international community.

Spain

- EU Emissions Trading (ETS) scheme
- Financial institutions will publish specific decarbonization targets for their lending and investment portfolio in line with the Paris Agreement as of 2023.
- Tax on HFC gases bases in 15 €/TCO_2-eq for all gases placed in the market

United States

The Department of Energy will invest up to USD 24 million for research into technology that captures carbon emissions directly from the air, replicating the way plants and trees absorb carbon dioxide.

4.2.5.8 Cross-cutting/other - labour market and other

Canada

- Supporting sustainable jobs for workers: The Government provides support to workers who need to advance their skills to transition toward a net-zero world.

The Netherlands

- For each climate sector, a list is drawn up of jobs that are essential to achieve climate change objectives. The number of climate-relevant vacancies is monitored to quantify the tension in the labour market.
- The training and lifelong development of workers and job seekers are essential to have a sufficient number of well-trained people who can contribute to the energy transition.

Pakistan

- Investments in Nature Based Solutions
- Global Methane Pledge

Spain

- The training and lifelong development of workers and job seekers are essential to have a sufficient number of well-trained people who can contribute to the energy transition.
- Circular Economy: Several new regulations and plans for all sectors
- Carbon footprint national registry. Registry for sink projects. National compensation registry

4.2.6 Additional examples

269. The G20 Data Gaps Initiative 3[25] identified additional examples that require comprehensive robust, and comparable data to develop the right policy measures for the transition towards a low carbon economy. The current Paris Agreement emission targets demand that economies will undergo structural reforms, in particular within the energy sector.

270. To monitor the progress of such economic reforms, and also their impacts, it is advisable to compile SEEA air emission accounts (see Annex I in EU regulation 691/2011[26]), including the carbon footprints, which are consistent with macroeconomic indicators such as industrial output, value added and real Gross Domestic Product.

271. The SEEA air emission accounts are a necessary complement to the IPCC-based GHG emission inventories. A New Zealand example is presented in Figure **4.4** (SEEA: residency principle and perspectives of both producers and consumers; IPCC: territory principle and only production perspective).

Figure 4.4
Greenhouse gas emissions; production and consumption perspective; territory and residency principle

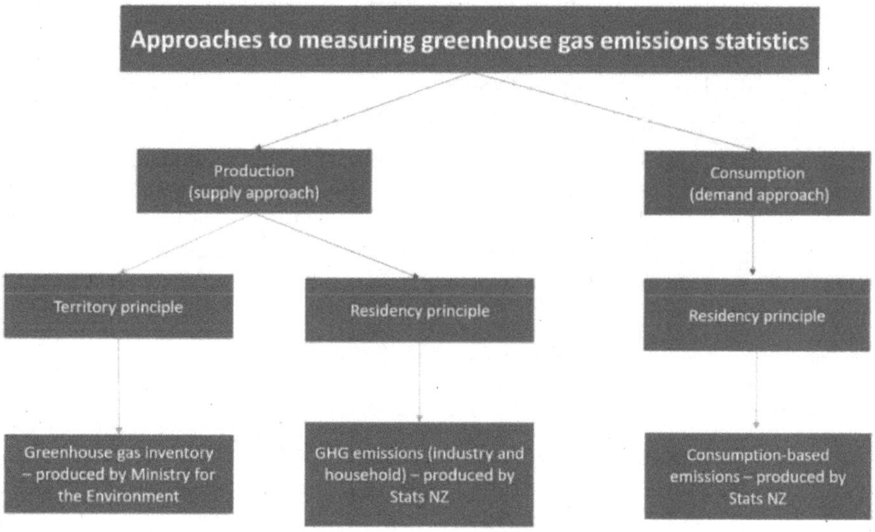

***Source*: Statistics New Zealand (2020).**

272. The air emission accounts not only provide GHG emissions from the perspective of producers, but also the consumption perspective (like carbon footprints). Both perspectives are essential to inform national mitigation policies. For example, the policies to reduce the carbon footprint in service-dominated economies will decrease the imports, and associated GHG emissions, from export-

[25] See more information about the G20 Data Gaps Initiative 3 in section 8.2.3.3 and at https://www.imf.org/-/media/Files/News/Seminars/DGI/Home/g20-dgi-3-workplan-people-planet-economy.ashx
[26] See https://eur-lex.europa.eu/legal-content/EN/TXT/PDF/?uri=CELEX:02011R0691-20220220&qid=1684238019231

oriented carbon intensive industries. Conversely, reducing the emission intensities of the energy sector and the carbon intensive industries also helps to reduce the national carbon footprints.

273. The consumption approach builds upon the residence-based emission estimates. It seeks to measure the emissions 'embodied' in a good or service throughout the entire supply chain required to produce that good or service for final use, so it requires the final user (household, business, or government) to be identified. This requires a link between industry supply of products and final user to be established, which requires the use of the industry-based production estimates.

Table 4.1
Comparison between Statistics Canada's National Inventory Report and Physical Flow Account

	National Inventory Report	Physical Flow Account
Mandate	The National Inventory Report (NIR) is Canada's official GHG inventory and annual submission to the United Nations Framework Convention on Climate Change.	The physical flow account (PFA) integrates GHG emissions estimates with economic activity across industries and households. The PFA aligns with the Canadian System of Macroeconomic Accounts
GHGs	Carbon dioxide (CO_2), Methane (CH_4), Nitrous oxide (N_2O) and synthetic fluorinated gases.	Carbon dioxide (CO_2), Methane (CH_4), Nitrous oxide (N_2O); Excludes: synthetic fluorinated gases.
Inclusion principle	**Territory principle:** emissions and removals occurring within Canadian Jurisdiction.	**Residency principle:** emissions and removals related to the operations of Canadian economic residents.
Aviation fuels	Includes emissions from foreign airlines flying over Canadian territory Excludes emissions from Canadian airlines flying over foreign territory.	Excludes emissions from foreign airlines flying over Canadian territory. Includes emissions from Canadian airlines flying over foreign territory.
Marine fuels	Includes GHG emissions from ships that travel between two ports located in Canadian waters. Excludes international voyages departing from or arriving in Canada.	Includes emissions from all Canadian ships for both domestic and international voyages. Excludes foreign ships.
Removals	Includes removals of CO2 from the atmosphere through: 1. its storage in woody vegetation on the landscape 2. its storage in soils of forest lands, croplands grasslands, and wetlands. 3. carbon capture, use, and storage technology.	Includes removals of CO2 from the atmosphere through carbon capture, use, and storage technology only.
Classification system(s)	The NIR's economic sector classifications do not conform to a standardized system such as the North American Industry Classification (NAICS) or the International Standard Industrial Classification (ISIC), although there are close similarities.	The StatCan's Input-Output Industry Classification (IOIC) is the classification used in the Supply and Use tables. It is a system used for classifying industries by codes similar to the North American Industry Classification System (NAICS).
Not included	The Energy sector excludes CO2 emissions from the combustion of biomass fuels (such as residential firewood, wood pellets, ethanol, and biodiesel).	Excludes emissions from solid and liquid waste. Excludes also emissions from non-energy products from solvent use.

Source: Canadian Centre for Energy Information (2024).

274. Understanding the source of GHG emissions is key in developing effective policy. Estimates of GHG emissions broken down by multinational and domestic enterprises would help policymakers better understand where GHG emissions are generated, who owns the enterprises generating them, and where the associated goods and services are consumed.

275. The Canadian example of complementary approaches (see Table **4.1**) shows that the scope of the official GHG inventory and SEEA approach differ. The National Inventory Report (guided by the COP/CMA decisions and reporting tables and their footnotes and the IPCC guidelines) considers biomass as carbon-neutral, due to its short carbon cycle, and therefore excludes from the energy sector the emissions from biomass[27], whereas the Physical Flow Account (SEEA) includes the biomass emissions linked to economic production and consumption activities by industries and households. Due to the residency principle, the international transport emissions from international flights by domestic airlines (and shipping companies) are included in the Physical Flow Account. However, IPCC still excludes such emissions, although the national inventories include bunker fuels as a memo item. The Canadian Physical Flow Accounts focus on the three main GHGs (CO_2, CH_4, and N_2O) and therefore do not include fluorinated gases, but they are conceptually included in the SEEA air emission accounts (e.g., see Annex I in EU regulation 691/2011).

276. Emission reduction policies related to international aviation are covered by the International Civil Aviation Organization (ICAO).[28] In 2022, the ICAO Assembly adopted, in a resolution[29], a collective long-term global aspirational goal for international aviation of net-zero carbon emissions by 2050, in support of the Paris Agreement's temperature goal. To achieve this goal ICAO is pursuing a basket of measures including aircraft technology improvements, operational improvements, sustainable aviation fuels, and market-based measures such as the Carbon Offsetting and Reduction Scheme for International Aviation (CORSIA).[30] To ensure a level playing field for sustainable air transport, the EU regulation 2023/2405[31] was adopted (in October 2023), as part of the RefuelEU aviation initiative.[32] This European obligation promotes the use of more renewable and low-carbon fuels by setting targets for the share of sustainable aviation fuels at Union airports (every five years from 2025 to 2050) to reduce the carbon footprint of the aviation sector.

277. Emission reduction policies related to international shipping are covered by the International Maritime Organization (IMO). In July 2023 the Marine Environment Protection Committee adopted a revised IMO strategy[33] to reduce the GHG emissions from international shipping towards net-zero around 2050. These emissions should be reduced, compared to 2008, by at least 20 per cent in 2030 and by at least 70 per cent in 2040. To achieve these goals, the energy efficiency of (new) ships will be improved, including the introduction of technologies, fuels and/or energy sources that emit zero or near-zero GHGs. As part of the FuelEU maritime initiative[34] there is a European obligation to increase the use of renewable and low carbon fuels (EU regulation 2023/1805[35]; adopted in September 2023). This sets targets (every five years from 2025 to 2050) for the reduction of GHG emissions per unit of energy.

---

[27] The net biomass emissions are captured in the LULUCF sector. However, the national LULUCF net emissions do not tell the full story due to international trade of, e.g., wood pellets. Biomass emissions are not included in the IPCC inventory of the country importing wood pellets; it is recorded in the LULUCF sector of the country exporting wood pellets (in case of net deforestation, it is recorded as a positive emission).

[28] https://www.icao.int/environmental-protection/pages/climate-change.aspx

[29] https://www.icao.int/environmental-protection/Documents/Assembly/Resolution_A41-21_Climate_change.pdf

[30] https://www.icao.int/environmental-protection/CORSIA/pages/default.aspx

[31] https://eur-lex.europa.eu/legal-content/EN/TXT/PDF/?uri=OJ:L_202302405&qid=1700055999672

[32] https://www.consilium.europa.eu/en/press/press-releases/2023/10/09/refueleu-aviation-initiative-council-adopts-new-law-to-decarbonise-the-aviation-sector/

[33] https://www.imo.org/en/MediaCentre/PressBriefings/pages/Revised-GHG-reduction-strategy-for-global-shipping-adopted-.aspx

[34] https://www.consilium.europa.eu/en/press/press-releases/2023/07/25/fueleu-maritime-initiative-council-adopts-new-law-to-decarbonise-the-maritime-sector/

[35] https://eur-lex.europa.eu/legal-content/EN/TXT/PDF/?uri=CELEX:32023R1805&qid=1700058191434

278. The GHG emission calculations for international aviation and shipping, related to ICAO and IMO net-zero targets, may use a life-cycle accounting approach, i.e., well-to-wing and well-to-wake, respectively. It covers the entire process of fuel production, delivery and use onboard, and all emissions produced therein. If the aviation and shipping sectors arrange emission reductions in the fuel production and delivery sectors, there is a risk that these emission reductions are counted twice: (1) in the emission monitoring of international aviation and shipping and (2) in the monitoring of the Enhanced Transparency Framework of the Paris Agreement. Such double counting may even get worse, if the aviation and shipping sector have their own global carbon markets (such as CORSIA) to trade their emission credits (see also Box 4.1).

279. To achieve mitigation targets it is important to share national good practices on policies and measures, especially as countries are often faced with the same hurdles and problems. To monitor, support, implement and evaluate mitigation policies, and accompanying measures, use can be made of the European environment information and observation network (Eionet),[36] in particular the Eionet Group on Climate Change Mitigation and Energy Systems.

280. The EEA briefing[37] on the prospects for decarbonization in the agriculture sector can be considered as a good practice. This contains an overview of most frequently reported measures like optimizing livestock diets, improving manure management systems and on-farm energy efficiency, and reducing the nitrogen quantity applied to soils. It also mentions the 'gaps' in reported measures like reducing livestock numbers, encouraging dietary change (fewer animal products), and supporting precision farming and sustainable soil management. Implementing circular economy actions across the full agri-food value chain can further reduce GHG emissions. In addition, farmers can increase carbon sequestration by applying specific agricultural practices (e.g., through agroforestry).

281. Carbon sequestration and storage by world's forests is part of the framework for reducing emissions from deforestation and forest degradation in developing countries, known as 'REDD+'[38] (see Figure 4.5). The '+' stands for additional forest-related activities that protect the climate, namely sustainable management of forests and the conservation and enhancement of forest carbon stocks. The SEEA ecosystem accounting can play a role here.

282. To inform long-term strategies, climate change research community has adopted a framework to facilitate the integrated analysis of future climate impacts, vulnerabilities, adaptation, and mitigation, including integrated, harmonized and consistent long-term scenarios such as Shared Socioeconomic Pathways (SSP)[39] and Representative Concentration Pathways (RCP).[40] These future trajectory projections are built upon historical data, which can be provided by NSOs. This includes statistics on, e.g., economic growth, energy use and supply, emissions, demographics, urbanization, and land use. The long-term strategies and scenarios both benefit from the high statistical quality requirements that apply to NSOs.

---

[36] https://www.eea.europa.eu/en/about/who-we-are/our-knowledge-network-eionet

[37] https://www.eea.europa.eu/publications/Progress-and-prospects-for-decarbonisation/progress-and-prospects-for-decarbonisation

[38] https://unfccc.int/topics/land-use/workstreams/redd/what-is-redd

[39] https://tntcat.iiasa.ac.at/SspDb/dsd?Action=htmlpage&page=10#intro

[40] https://tntcat.iiasa.ac.at/RcpDb/dsd?Action=htmlpage&page=welcome. See also Ingole, V. et al. (2022).

Figure 4.5
Geographic distribution of 377 REDD+ projects (2018-2020): avoided deforestation and degradation and forest conservation (AD), afforestation/reforestation/revegetation (AR) and other goals (other)

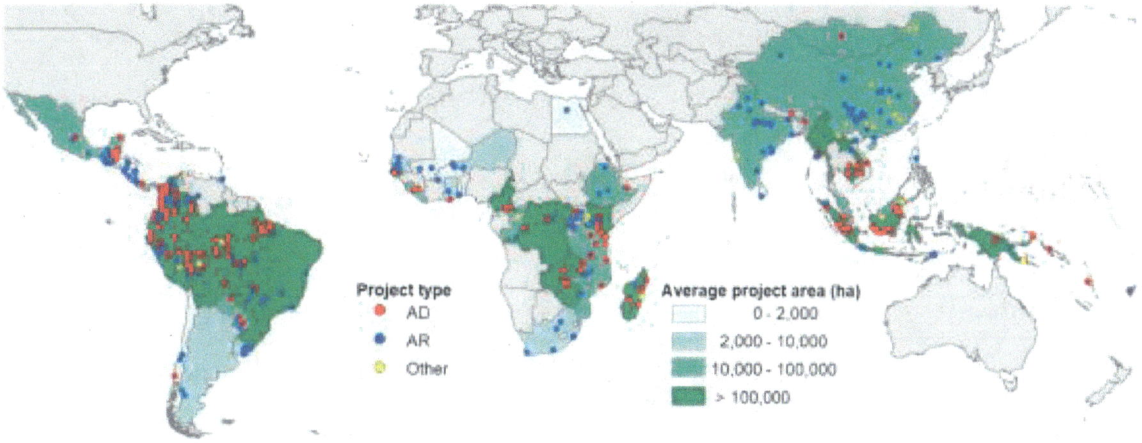

Source: Atmadja, S. et al. (2022).

4.3 How NSOs can contribute

283. The main role of NSOs is to provide the necessary data to policymakers and other data users. This asks for basic data collection, enriched by microdata-linking (if needed), developing new statistics, and disseminating all statistics on a regular basis and in an easily accessible way. In order to compare the statistics country-by-country, indicator sets are established at an international level, using consistent harmonized statistical frameworks, like IPCC and SEEA, and taking into account relevant official statistics (e.g., on energy, agriculture and transport).

284. In the process of providing data for climate change mitigation, clear institutional arrangements and coordination are essential as the role of data producer is not only played by NSOs, but also by other governmental agencies and non-governmental actors. Depending on national circumstances and NSO mandate, NSO can play a bigger role than just providing data. For instance, it can play a role in a coordination mechanism with clear definition of roles and responsibilities, ensure that all stakeholders are onboard while data collection is efficient and fit-for-purpose. For further discussion see section 2.2 (Role of the NSO), sections 3.3, 4.3, 5.3, 6.3, 7.3, and 8.3 (How NSOs can contribute; in different areas), section 9.1 (Governance and coordination), section 9.2 (Engagement with policymakers), and section 9.5 (Collaboration with researchers and academia).

285. Section 4.3 is set up as follows. Using an international perspective and based on the mitigation definition in section 4.2.1, the indicators and dashboards related to the mitigation policy domain are described. This is enriched by showing some clarifying country examples. The importance of energy statistics is stressed. Other official statistics important for the mitigation policy domain are also highlighted.

286. The Eurostat Statistics Explained website presents various statistical topics in an easily understandable way (Eurostat, n.d.). It includes articles on GHG emissions (inventories, air emission accounts and carbon footprints), on climate change - driving forces, and on SDG 13 on climate action. Much of the data presented comes from SEEA-CF, which is usually compiled by NSOs and can be used as a consistent framework to assess policy responses to climate change (Statistics Netherlands, 2010). As a result of an upcoming delegated act by the EU Commission, in near future the EU SEEA data may also cover data on climate change mitigation investments as part of environmental goods and services sector accounts.

Figure 4.6
Net greenhouse gas emissions EU 1990-2021 (including international aviation and LULUCF).

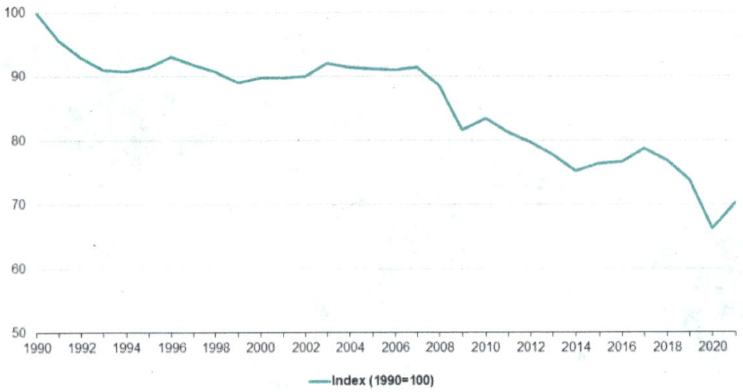

***Source*: Eurostat (n.d.), Emission inventories.**

287. Section 4.3 shows what data the statistical community can offer to the mitigation policy domain using a set of figures illustrating the text on indicators and dashboards at in Section 4.3.1:

(a) Figure 4.6: Net GHG emissions; EU 1990-2021 (including international aviation and LULUCF);

Figure 4.7

(b) Greenhouse gas emissions by source sector; EU 2020;

Figure 4.8

(c) Gross electricity and heat production from renewables and biofuels; EU 1990-2020;

(d) Figure 4.9: Development of GHG emissions compared to GDP and population; EU 1990-2020;

Figure 4.10

(e) Greenhouse gas emissions by economic activity; EU 2008 and 2021;

(f)

(g) Figure **4.11**: CO_2 emissions - production and consumption perspective; EU, 2019;

(h) Figure 4.12: Net GHG emissions from land use and forestry; EU and by country, 2016 and 2021;

(i) Figure 4.13: Quarterly GHG emissions (seasonally adjusted); 2010-2020 for (a) Advanced Economies and (b) Emerging and Developing Economies;

(j) Figure 4.14: Two EU Green Deal indicators (2014-2020): (a) Share of zero-emission vehicles and (b) Premature deaths due to exposure to fine particulate matter;

(k) Figure 4.15: Research, development and demonstration (RD&D) budgets on renewable energy and other low-carbon technologies, as % of total public energy RD&D, by country, 2021 (or latest year available);

288. Figure **4.16** - Figure **4.21** show examples of data on selected country examples described in Section 4.3.2:

(a) Figure **4.16** GHG emissions (Luxembourg, 2008-2021); territory approach (inventories) versus residential approach (SEEA);

(b) Figure **4.17** GHG emissions (New Zealand, 2017); flow from production to consumption;

(c) Figure **4.18** and Figure **4.19**: Footprint emissions; domestically produced and embodied in transport (Rest of the World): flow from production to consumption (Denmark; 2021) and climate footprint (Denmark; 1990-2021);

(d) Figure 4.20: Decomposition analysis: underlying drivers of the development of CO_2 emissions (the Netherlands, 1971-2020);

(e) Figure 4.21: Levels of monitoring of climate change monitoring policies (the Netherlands).

289. Section 4.3.3, which emphasizes the importance of energy statistics and the role of NSOs, includes Figure 4.22: From primary energy data to energy-related climate indicators for policy making.

Figure 4.7
Greenhouse gas emissions by source sector; EU 2020

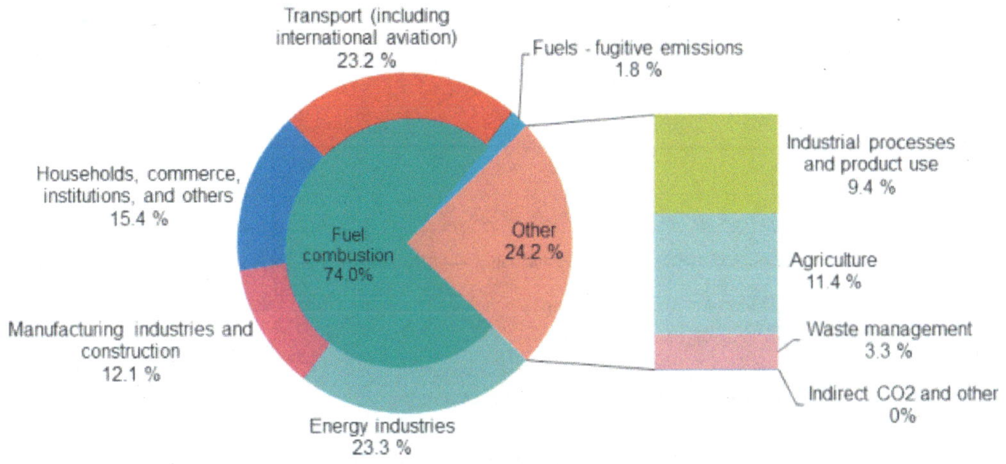

Source: Eurostat (n.d.), Climate change - driving forces

Figure 4.8
Gross electricity and heat production from renewables and biofuels; EU 1990-2020

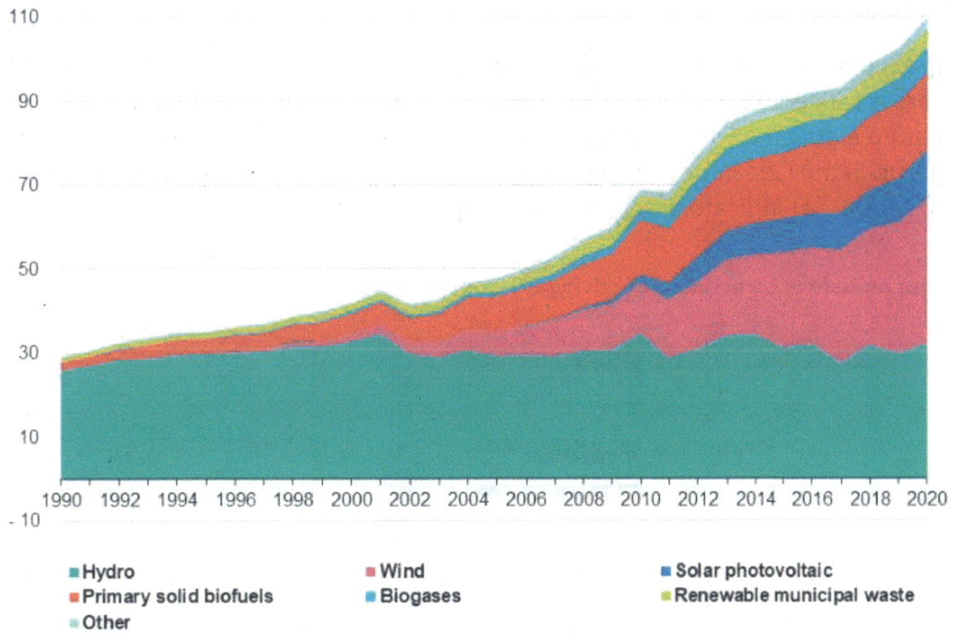

Source: Eurostat (n.d.), climate change - driving forces

Figure 4.9
Development of Greenhouse gas emissions compared to Gross Domestic Product (GDP; economic growth) and population; EU 1990-2020

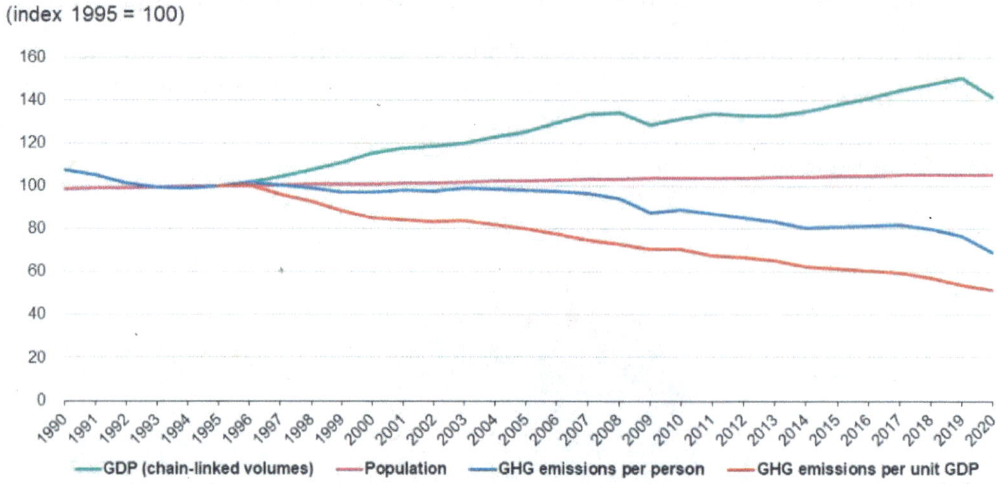

Notes: GHG emissions total (excluding LULUCF - Land use, land use change and forestry - and memo items, including international aviation)
Data on EU-27 GDP only available for 1995 onwards.

Source: Eurostat (n.d.), climate change - driving forces

4.3.1 Indicators and dashboards at international level

290. Two official international indicator sets exist for countries to use when developing their own indicator sets for climate change-related statistics. In 2020, the CES indicator set was endorsed (see section 2.3.6) and contains 44 indicators divided in 5 areas (drivers 9; emissions 9; mitigation 8; impacts 13; adaptation 5).

291. In 2022, the UNSD Global Set was adopted (see section 2.3.7) and contains 158 indicators divided in 5 areas (drivers 26; mitigation 18; impacts 54; adaptation 32; vulnerability 28)[41]. The UNSD Global Set has an extra area (vulnerability), whereas the drivers area also contains emissions (which is a separate group in the CES indicator set). Suriname recently made an extensive inventory of the data availability of Global Set indicators, see Suriname report (August 2023).[42]

Figure 4.10
Greenhouse gas emissions by economic activity; EU 2008 and 2021

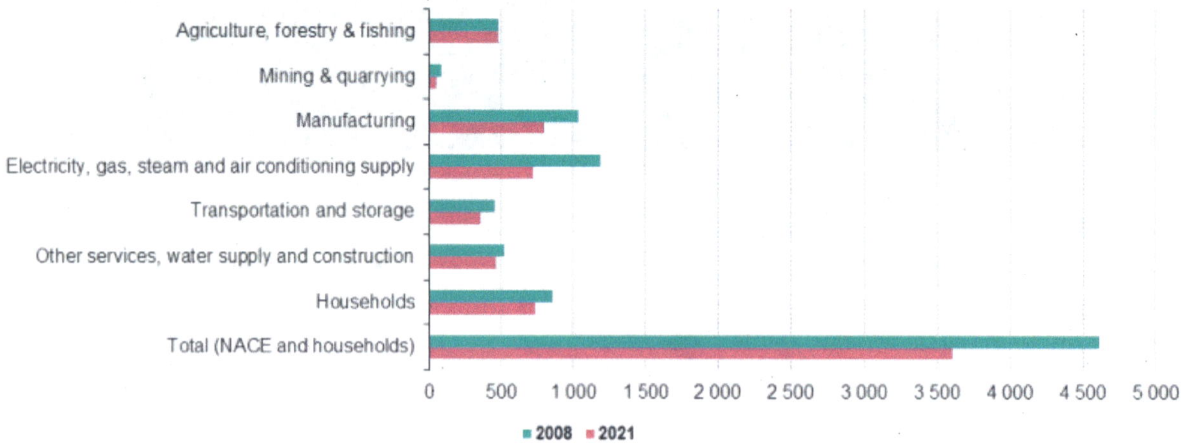

Source: Eurostat (n.d.), air emissions accounts

[41] Many of the 158 Global Set indicators belong to more than one of the five groups. If multiple belongings are considered than the total number of indicators applicable to mitigation is 28.
[42] https://statistics-suriname.org/wp-content/uploads/2023/08/First-Suriname-CC-report_21-aug23-Climate-Change-Statistics-and-Indicators.pdf

292. For the mitigation policy domain only the drivers, emissions and mitigation areas are essential (CES set: 26 indicators; Global Set: 44 indicators). The Eurostat Climate Change Database [43] distinguishes the same areas, although the presented data sets do not have a one-to-one relation with the CES indicator set and/or UNSD Global Set. The Eurostat Climate Change Database shows how many EU datasets are available for these areas. The emissions area contains 9 data sets, the drivers area 45 data sets (Energy 14; Transport 9; Industry 3; Waste 2; Agriculture 7; LULUCF 10) and the mitigation area 12 data sets. Thus, the Eurostat Climate Change Database already contains 66 data sets, with harmonized EU Member State data, on drivers, emissions and mitigation aspects.

293. The data in the Eurostat Statistics Explained articles (as shown in Figure 4.6 – Figure 4.12), combined with the data in the Eurostat Climate Change Database, show a significant overlap with the CES indicator set and the UNSD Global Set. In general, for the mitigation policy domain, the indicators from both sets in the emissions and drivers areas can be provided by already existing official statistics and inventory data. If more detailed or new mitigation policy data is needed, linked to the indicators in the emissions and drivers areas, this can be provided by the inventory and/or statistical community following the normal procedures to improve their official data.

Figure 4.11
CO_2 emissions – production and consumption perspective, EU, 2019

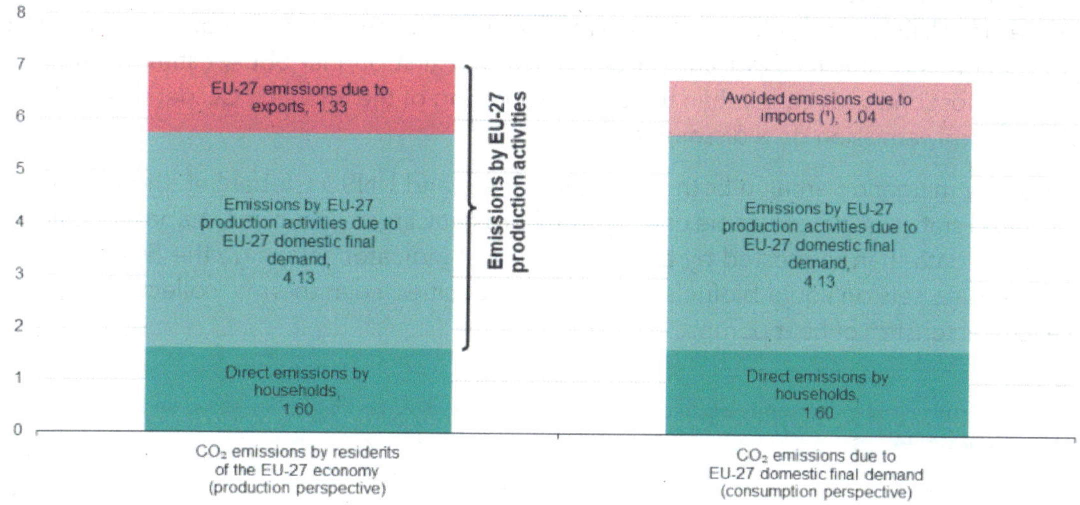

Notes: Estimates.
(¹) 'Avoided emissions due to imports' are based on the amount of carbon dioxide that would have been emitted in case the products imported would have been produced in the E-27 using EU-27 production technologies.

Source: **Eurostat (n.d.), carbon footprints**

43 https://ec.europa.eu/eurostat/web/climate-change/database

Figure 4.12
Net greenhouse gas emissions from land use and forestry, EU and by country, 2016 and 2021

(tonnes of CO_2 equivalent per km²)

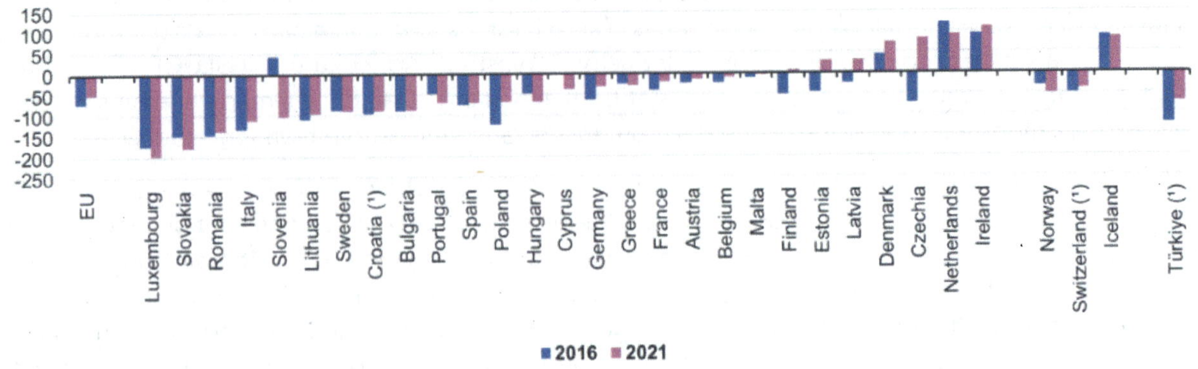

Note: Data for 2021 are provisional estimates based on the EEA approximated GHG inventory for the year 2021.
(¹) 2020 data (instead of 2021).

Source: **Eurostat (n.d.), SDG 13 - climate action**

294. One recent improvement is the quarterly dissemination of GHG emissions[44] (SEEA-based) by Eurostat, i.e., for all EU Member States at the same frequency as other macroeconomic reporting. This implies an improved timeliness of annual emissions: the first annual SEEA estimate (sum of four quarters) is almost one year earlier than the official reporting of the annual IPCC data. See also Figure 4.13 with quarterly emission data disseminated by IMF.

295. For the mitigation area, in both CES indicator set and UNSD Global Set, there is an overlap with the official statistics on renewable energy. The Eurostat Climate Change Database provides even more detailed data than requested by both international indicator sets, e.g., the Eurostat database also contains data sets on liquid biofuels production capacities, solar thermal collectors' surface, and technical characteristics of heat pumps.

Figure 4.13
Quarterly GHG emissions (seasonally adjusted), 2010-2020 in Advanced Economies (a) and Emerging and Developing Economies (b)

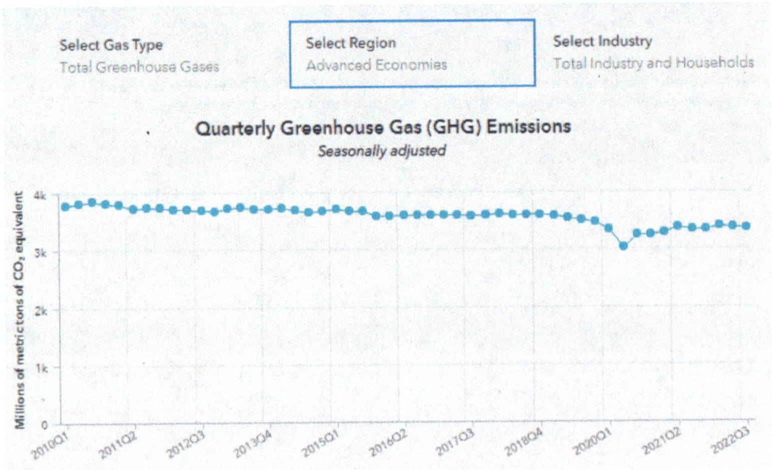

44 https://ec.europa.eu/eurostat/statistics-explained/index.php?title=Quarterly_greenhouse_gas_emissions_in_the_EU

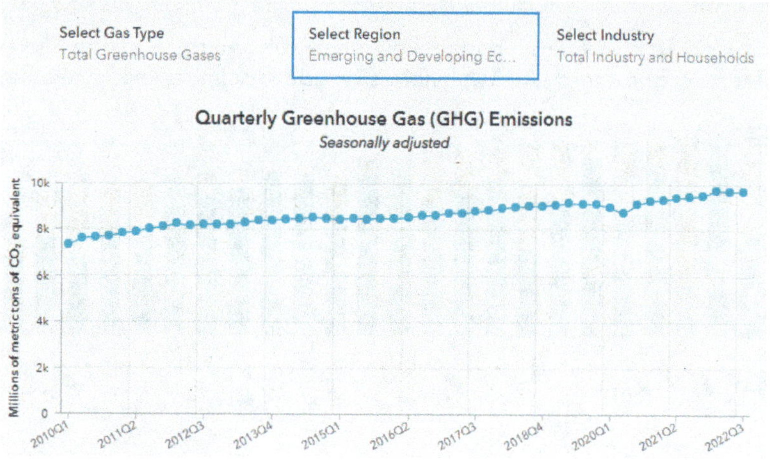

Source: IMF Climate Action Dashboard[45]

296. Several relevant indicators for the mitigation domain can also be found on the Eurostat Green Deal Dashboard. In addition to data on GHG emissions and renewable energy it also contains data on the share zero-emission vehicles and the premature deaths due to exposure to fine particulate matter (see Figure 4.14; EU total; data by Member State is also available). Reducing the consumption of fossil fuels by road traffic not only reduces the GHG emissions, but also improves the air quality and has positive health effects. In general, the inventory community uses in their compilation process the same activity data (e.g., amount of consumed fossil motor fuels and/or the number of travelled km's) for estimating the GHG emissions and for the emissions of air pollutants.

Figure 4.14
Share of zero-emission vehicles and premature deaths due to exposure to fine particulate matter (PM2.5: diameter less than 2.5 micrometer; rate: number per 100.000 people), 2014-2020.

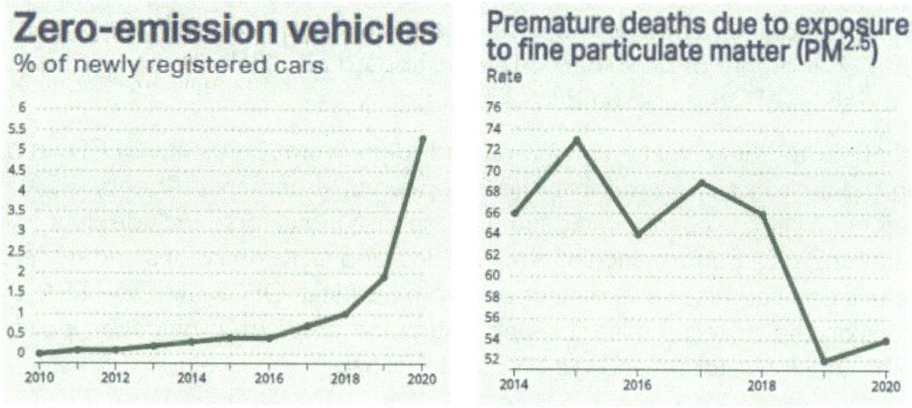

Source: Eurostat Green Deal Dashboard[46]

297. Worldwide data is provided by the OECD Climate Action Dashboard, as part of the International Programme for Action on Climate (IPAC), which can be used to track progress towards climate objectives and to provide a snapshot of country climate action. Similar to Eurostat data the OECD Climate Action Dashboard contains GHG emissions data (trends and trajectories; structure and intensities) and data on the change in the energy mix. The dashboard also includes data on policy adoption, climate-related pricing and taxation (e.g., effective carbon rates and climate related tax revenues), and climate related innovation (e.g., patent applications in climate change mitigation technologies and 'Research, Development, and Demonstration' budgets on renewable energy and other low-carbon technologies, see Figure 4.15).

45 https://climatedata.imf.org/
46 https://ec.europa.eu/eurostat/cache/egd-statistics/

Figure 4.15
Research and Development Demonstration budgets on renewable energy and other low-carbon technologies as % of total public energy RD&D, by country, 2022 (or latest year available).

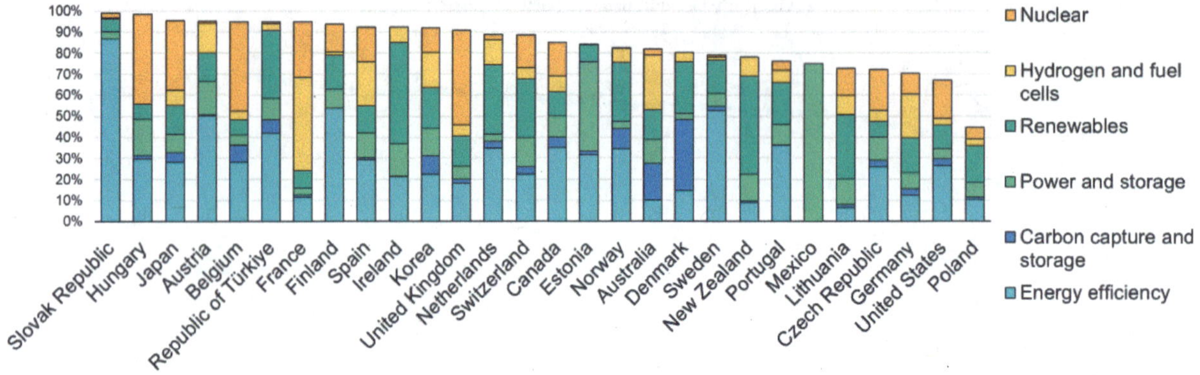

***Source*: Based on IEA data from IEA (2023) Energy Technology RD&D Budgets**[47]

298. Worldwide data is also provided by the IMF Climate Action Dashboard[48], which can be used as a tool linking climate considerations and global economic indicators. Similar to Eurostat data and the OECD Climate Action Dashboard, it contains the same kind of GHG emissions data (also quarterly) and data on electricity production (renewable and non-renewable). The regional data includes not only country data, but also a breakdown of global data by 'advanced economies' and 'emerging and developing economies' (see Figure **4.13**; the Covid dip is clearly visible at 2020-Q2, and the trends are opposite). The IMF Climate Action Dashboard also covers cross border indicators like CO_2 emissions embodied in gross imports and exports (at country level), the (bilateral) trade in low carbon technology products, and CO_2 emissions by firm ownership (domestic controlled enterprises versus foreign controlled multinational enterprises). Other IMF dashboard indicators are focused on government policy (e.g., environmental taxes and fossil fuel subsidies; SDG 12.1.c, UNEP (2019)) and on climate finance (e.g. CO_2 emissions in gross fixed capital formation of direct investment, green bonds, and carbon footprint of bank loans). The dashboard also includes a world map showing for each country the share of forest area (and its carbon stock).

299. In addition to the climate change-related data from Eurostat, OECD and IMF, other dissemination channels exist. For example, the European Environmental Agency (EEA) disseminates various indicators on climate change mitigation. These EEA indicators[49] include data on atmospheric GHG concentrations, CO_2 performance of new vehicles, and GHG emissions under the EU Emissions Trading System. If one combines all indicator sets from Eurostat, OECD, IMF and EEA, they cover the same indicators on the drivers, emissions, and mitigation areas (the mitigation policy domain) as presented in the CES indicator set and the UNSD Global Set.

300. The energy sector accounts for about three quarters of total GHG emissions globally, hence addressing the clean energy transition is central in achieving climate mitigation objectives. The IEA-CMCC Weather for Energy Tracker[50] and IEA-OECD Climate Hazard Exposure Tracker[51] are innovative data focused platforms with weather variables and climate indicators with different time and geographical granularities, ranging from 1979 to the latest available month. Given the strong interlinkage between energy and weather, detailed weather data can be used to inform infrastructure planning in the form of solar, wind and hydropower potentials for electricity generation, improving the effectiveness of such installations. Additionally, cooling and heating degree days (indicators derived from temperature data) are directly linked to energy demand for cooling and heating purposes. Long, country-level time series can provide essential understanding of energy demand

[47] http://www.iea.org/statistics
[48] https://climatedata.imf.org/
[49] https://www.eea.europa.eu/ims
[50] https://www.iea.org/data-and-statistics/data-tools/weather-for-energy-tracker
[51] https://www.iea.org/data-and-statistics/data-tools/weather-for-energy-tracker?tab=Climate+Hazard+Exposure+Tracker

needs when dimensioning future energy systems, directly tied to clean energy transitions and climate change mitigation. The IEA-OECD Climate Hazard Exposure Tracker explores indicators beyond the energy landscape, providing insight for climate hazards such as droughts, wildfires, cyclones, storms and flooding directly related to agriculture, other various economic activities and people's qualities of life. Such data helps analysis on how these extreme weather events affect communities and facilitate planning for their occurrence and related consequences.

4.3.2 Country examples

301. The international dashboards and other data portals do not show mitigation related data from all the countries around the world. The countries differ in their capacity to provide climate change-related statistics. It ranges from basic, intermediate to advance, with each of them having their own roadmap to develop climate change-related statistics (see UNECE, 2017).

Armenia

One country example is Armenia's 2020 roadmap[52]. This roadmap describes how the GHG inventory from Armenia is organized, including its legislative basis. It further sets priorities on the development of climate change-related statistics like GHG emissions (and its sources) and mitigation statistics (e.g., installation of solar, wind and geothermal power). Armenia's roadmap contains recommendations such as more involvement of the statistical office of Armenia (Armstat) in their inventory system (e.g., active participation in data collection and quality assurance), sufficient capacity building and well-trained NSO staff, improved access to (new) administrative data, more use of geographic information systems (GIS), and a further development of SEEA air emission accounts.

302. The emissions from inventories differ from the ones in the SEEA air emission accounts, i.e., territory versus residence principle (see Figure 4.16; Luxembourg). The territory approach (inventories) includes emissions from motor fuels purchased in Luxembourg. However, a significant portion is not purchased by Luxembourg residents due to a high cross-border commuting by non-residents. This non-residential part are subtracted from the inventory emissions to obtain the emissions according to the residence principle. On the other hand, the emissions from domestic airlines are added, i.e., they are not accounted for in the inventories, whereas they are included according to SEEA air emission accounts.

Luxembourg

Figure 4.16
Greenhouse gas emissions (Luxembourg, 2018-2021); territory approach (inventories) versus residential approach (SEEA)

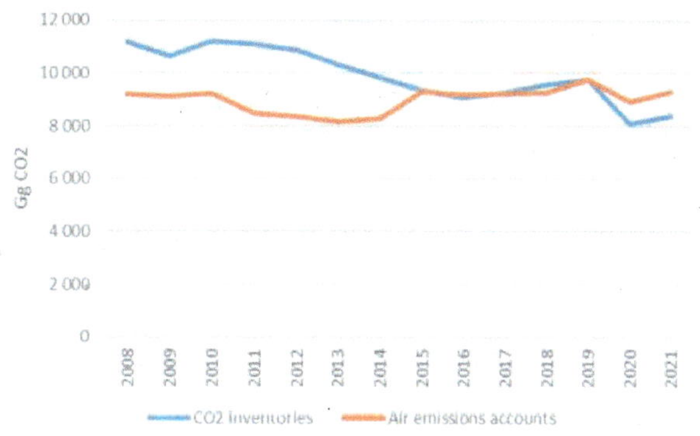

Source: Thunus, O. (2023).

303. The SEEA is used to monitor the decarbonization of national economies, e.g. by comparing the emission intensities by sector. In similar sense, SEEA sectoral emissions can be compared with

[52] https://www.armstat.am/file/doc/99525613.pdf

received tax revenues for each sector. Another advantage of SEEA is its coverage of both domestic production and domestic consumption (see Figure 4.17; New Zealand).

New Zealand

In New Zealand the emissions from final domestic consumption are one quarter smaller than the domestic production emissions due to the significant amount of emissions from dairy manufacturing for non-domestic consumption.

Figure 4.17
Greenhouse gas emissions (New Zealand, 2017); flow from production to consumption

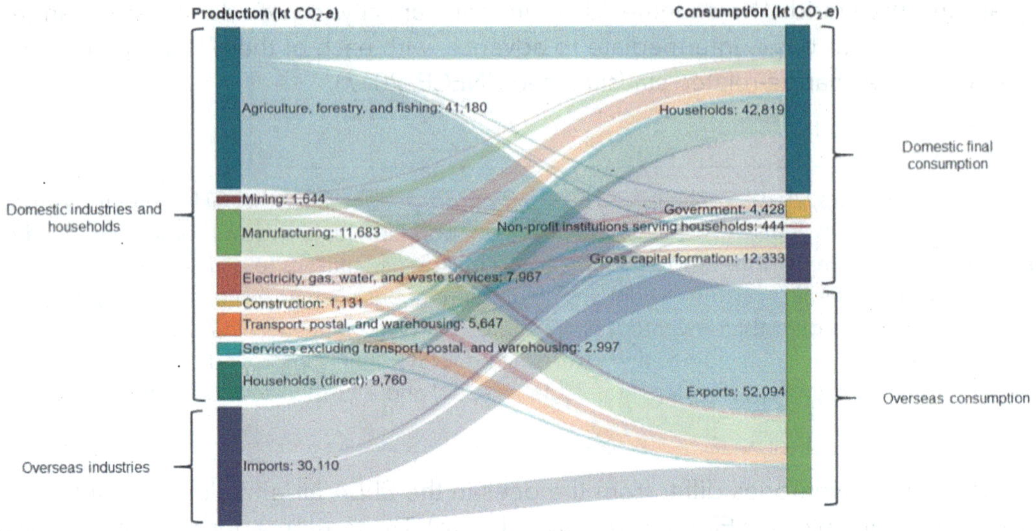

Source: Statistics New Zealand (2020).

304. The footprints from Denmark and New Zealand show a similar pattern: the footprint emissions are lower than the domestically produced emissions (see Figure 4.18 and Figure 4.19; Denmark).

Denmark

In 2021 the Danish climate footprint was 21 per cent lower than in 1990. This decrease is primarily due to a decrease in emissions that take place in Denmark, while the emissions abroad remain relatively constant over time. Mitigation policies focusing on footprints may not only result in lower domestically produced emissions, but they may also put extra pressure on reducing the emissions of other countries, i.e., to reduce the emissions embodied in their exports.

Figure 4.18
Connection between production and consumption-based emissions in Denmark (Denmark; 2021; in megaton CO_2-equivalent)

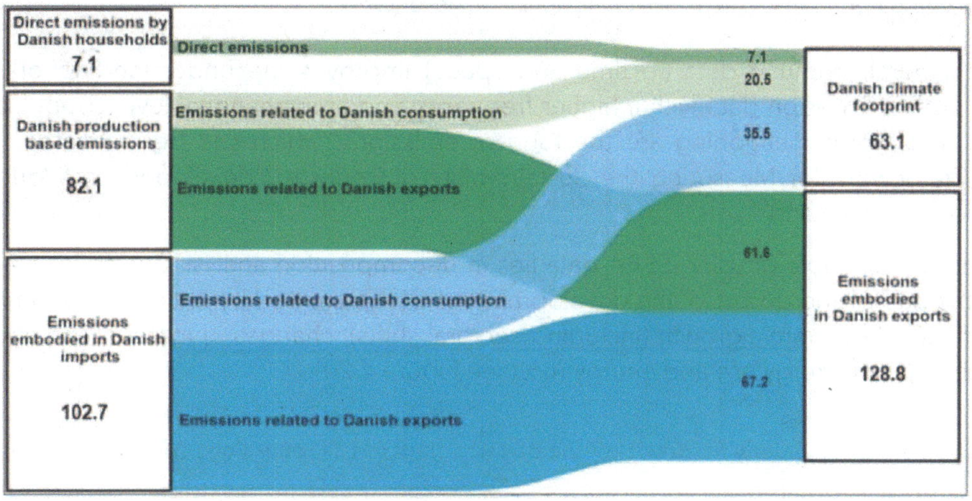

Source: Statistics Denmark, Jensen, P. and B. Iliev (2022).
Note: This is labelled as "experimental statistics" and may be subject to future revision

Figure 4.19
Climate footprint (Denmark; 1990-2021); domestically produced and embodied in imports

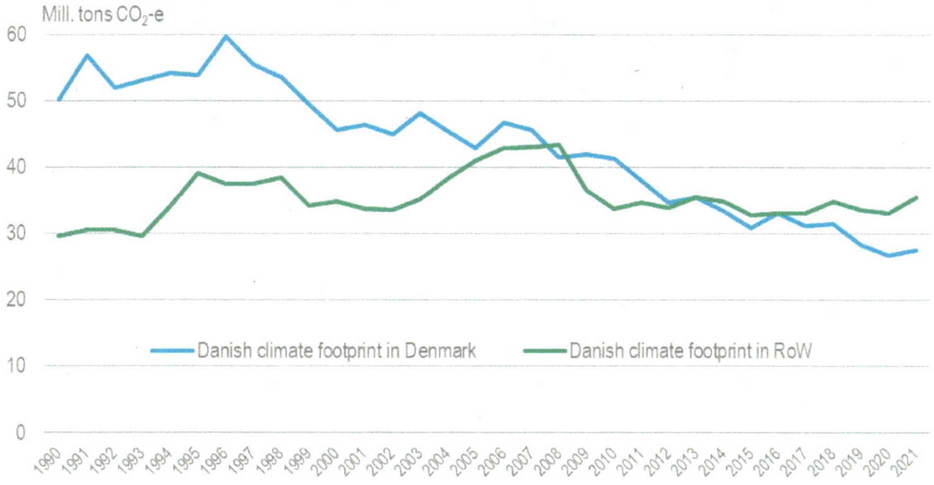

Source: Statistics Denmark, Jensen, P. and B. Iliev (2022).
Note: This is labelled as "experimental statistics" and may be subject to future revision

305. In general, footprint estimates are on an experimental basis since there are no international standards or reporting requirements for the compilations of consumption emissions. Moreover, to estimate emissions embodied in imports based on the production technologies used abroad, a global model and accompanying dataset including all inter-industry trade flows are needed. Such standardized global dataset is currently not available, although Eurostat is producing an EU-based one (FIGARO tables[53]) that is updated on an annual basis.

The Netherlands

In 2020 the Dutch Ministry of Economic Affairs and Climate Policy requested Statistics Netherlands to develop IPCC quarterly GHG emissions[54], complementary to the already existing Dutch SEEA quarterly emissions (since

[53] https://ec.europa.eu/eurostat/web/esa-supply-use-input-tables/figaro
[54] https://www.cbs.nl/-/media/_pdf/2020/37/ipcc-eng.pdf

2010). Both IPCC and SEEA perspectives are presented in a quarterly news release to be disseminated two and a half months after the end of the quarter. This implies that the first annual emission estimate is available 75 days (!) after the reporting year, see NL news release on 2022 GHG emissions[55] (March 14, 2023). A huge improvement in timeliness compared to the regular GHG emissions from inventories.

306. In general, granular data (in time and space) improves the understanding of a nation's emissions profile. Emission data with a higher frequency and at subnational levels both lead to new insights[56]. Also worth mentioning are the tourism emissions, which can be estimated within the Statistical Framework for Measuring the Sustainability of Tourism[57] developed by World Tourism Organization (UN Tourism)[58].

307. Another example of using SEEA data lies in decomposition analysis, which is useful to gain insight in the underlying drivers of the development of GHGs emitted by national economies, i.e., to quantify effects of economic growth (including sectoral shifts), changes in material use and material mix, and changes in energy use and energy mix (see Figure 4.20).

Figure 4.20
Decomposition analysis: underlying drivers of the development of CO_2 emissions (The Netherlands, 1971-2020)

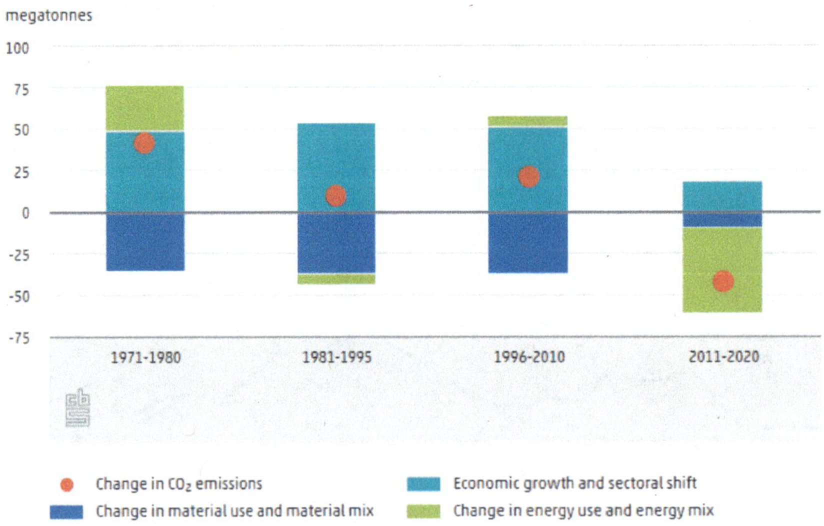

***Source*: Index Decomposition Analysis, Statistics Netherlands[59]**

The Netherlands

Until 2010 the economic growth was the driving force for the increase in CO_2 emissions in the Netherlands, whereas more efficient use of (raw) materials and recycling had a mitigating effect on CO_2 emissions in the same period. The last decade (2011-2020) mainly energy savings and the increase in renewable energy played a role in the decline of CO_2 emissions. Similar to the arguments used for footprint estimates, decomposition analysis results are generally on an experimental basis. Further testing of the robustness of the model is needed.

308. An energy transition is needed to achieve the reduction targets (2030 and 2050) for GHG emissions. The energy infrastructure must be such that supply, and demand are in balance when energy production becomes more local rather than centralized. This not only includes transitions in

[55] https://www.cbs.nl/en-gb/news/2023/11/greenhouse-gas-emissions-9-percent-lower-in-2022
[56] For example, see the news releases from Statistics New Zealand – quarterly emission estimates https://www.stats.govt.nz/news/total-greenhouse-gas-emissions-decrease-0-5-percent-in-the-december-2023-quarter/) and by region (https://www.stats.govt.nz/information-releases/greenhouse-gas-emissions-by-region-industry-and-household-year-ended-2022)
[57] https://www.unwto.org/tourism-statistics/measuring-sustainability-tourism
[58] See a New Zealand pilot study on GHG emissions from tourism https://webunwto.s3.eu-west-1.amazonaws.com/s3fs-public/2022-12/symposium_sessC_NZ.pdf
[59] https://www.cbs.nl/en-gb/news/2023/47/material-efficiency-not-enough-to-reduce-co2-emissions

the energy supply, but also for the energy use in sectors like transport (electrification and biofuels) and the use of energy in the built environment. Housing characteristics can be monitored using censuses.

309. The 2021population census from Cyprus[60] included questions on type of heating, use of solar energy and existence of air condition facilities. The 2021 Greek population census[61] included questions on thermal insulation and the main sources for heating, cooling, cooking and hot water production. Also the 2022 population census in Serbia[62] included a question on type of heating. The 2022 population census in Ireland[63] showed that 23 per cent of the Irish households use renewable energy sources; over 70 per cent of the dwellings built since 2016 used at least one renewable energy source, over 40 per cent of this same cohort had heat pumps and approximately 30 per cent had solar panels.

Figure 4.21
Levels of monitoring of climate change mitigation policies

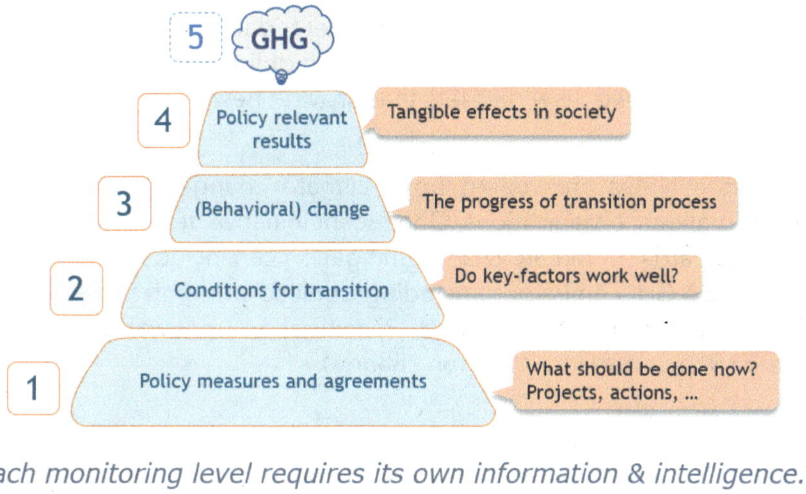

Source: **Statistics Netherlands**

310. Monitoring national mitigation policies should not only focus on reducing GHG emissions, but also on all kinds of aspects that enable a successful energy transition and related transitions in industry, agriculture, transport, and built environment (see Figure **4.21**; monitoring pyramid as used in the Netherlands). Just like the reduction of GHG emissions (5th level) also the tangible effects that can be seen in society (4th level; e.g., local charging stations for electric/hybrid cars) must be monitored. In addition, the three lower monitoring levels are particularly important for long-term innovation projects (climate investments) and other enablers for successful transition (e.g., training of heat pump installers), since it takes some time before their effects on the ultimate policy objectives at the top levels are visible.

311. Also perception, attitudes and behaviour of households, companies and other stakeholders can be monitored, e.g., by extending the already existing sentiment surveys to compose forward looking indicators in the climate change domain (see also Recommendation 5 of the G20 Data Gaps Initiative 3 in Table **8.1**). Another example is the news release on the share of Dutch people who are concerned about the impact of climate change, including their attitudes on other energy transition aspects[64].

---

[60] https://www.census2021.cystat.gov.cy/images/Census 2021 QST-EN .pdf
[61] https://www.statistics.gr/en/2021-census-pop-hous-results
[62] https://popis2022.stat.gov.rs/media/31011/016a-p2-engleski.pdf
[63] https://www.cso.ie/en/releasesandpublications/ep/p-cpsr/censusofpopulation2022-summaryresults/dwellingcharacteristics/
[64] https://www.cbs.nl/en-gb/news/2021/22/three-quarters-of-dutch-concerned-about-impact-of-climate-change

312. Another enabler for successful transition is a timely availability of critical minerals, e.g., cobalt, lithium, and rare earth elements) for appliances like batteries, solar panels and electric cars. Solar photovoltaic plants, wind farms and electric vehicles generally require more critical minerals to build than their fossil fuel-based counterparts (IEA, 2021). Improved statistics on critical minerals are certainly welcome, preferably based on consistent methodologies such as those in Material Flow Accounts (see section "Material flow accounts statistics - material footprints" in Eurostat, n.d.). Indicators can, for example, focus on material import dependency and self-sufficiency; not only for critical minerals, but also for other essential raw materials).

313. As official data provider the national statistical organizations play an eminent role in producing and disseminating statistics that are useful in a successful monitoring of climate change-related transition processes. The NSOs can do much more than simply providing SEEA data and official statistics on energy, industry, agriculture, transport and built environment, which is clearly shown in two papers from Ireland on data stewardship in the environment domain (Morrin, P., 2022).and on the use of administrative microdata for climate-related statistics (Brady, G., 2022). The latter paper describes Irish examples of microdata use, e.g., utility meter data (electricity, gas and water), building energy ratings, and vehicle odometers. This paper also highlights the unique legal role of NSOs to combine microdata files to create new datasets that provide new insights to policymakers and other data users.[65]

314. More and more NSOs are setting up a climate change statistics portal, analogous to dashboards from international organizations. One recent initiative, from Spain, is shown in Box **4.2**. In general, national dashboards on climate change mitigation can include information on (1) emissions (trends in IPCC and SEEA GHG emissions; including detailed bridging tables[66], (2) drivers (trends in energy mix and sectoral energy intensity), and (3) mitigation (renewable energy, building energy ratings, public energy RD&D, and forest carbon change).

[65] For more information, see the web page of the joint UNECE/IEA/Eurostat Webinar on Administrative Microdata for Climate Change, Energy and Environment Statistics, which took place on 30 March 2023: https://unece.org/statistics/events/MicrodataWebinar2023

[66] See the United Kingdom example: https://www.ons.gov.uk/economy/environmentalaccounts/datasets/ukenvironmentalaccountsatmosphericemissionsbridgingtables

Box 4.2
Dashboard of environmental indicators in the framework of the European Green Deal published by INE Spain.

INE Spain publishes a new dashboard of environmental indicators in the framework of the European Green Deal

In recent years, the National Institute of Statistics of Spain (INE Spain) has developed seven new environmental accounts covering various areas, such as emissions to the atmosphere, material flows, environmental taxes, waste, environmental expenditures, energy and environmental goods and services. These accounts complement the traditional environmental statistics of the INE, focused on the generation and treatment of waste, water and environmental protection expenditures.

These statistics and accounts provide a comprehensive set of environmental indicators relevant to users and decision-making in the environmental field.

However, it is important to underline that it is difficult for users to obtain an overview that allows them to synthetically analyse the situation and evolution of the different indicators published by the INE in areas such as waste, energy, emissions, consumption, materials productivity, environmental taxation, growth and green employment, among others.

Europe is considered a global leader in the transition towards the circular economy and the fight against climate change. The European Green Deal, presented by the European Commission in 2019, aims to achieve climate neutrality by 2050. This comprehensive strategy includes actions related to clean energy, circular economy, sustainable mobility, sustainable agriculture or biodiversity.

Many of the indicators developed by the INE are aligned with the European Green Deal, the Spanish Circular Economy Strategy, Climate Change and the 2030 Agenda for Sustainable Development.

The aim of this new publication is to monitor these targets through indicators that reflect the national position compared to the European average.

Under the European Green Deal, an interactive dashboard is established showing the evolution of selected indicators and performance tables with time series. This scorecard is divided into thematic axes related to the main European environmental strategies, such as the circular economy, climate change and biodiversity. Filters are provided to personalize the consultation of the indicators, as well as descriptions of each indicator to facilitate their understanding.

The dissemination of information is being carried out through a website that includes both national and European Union data, allowing comparisons to be made. In addition, result tables with detailed metadata on the sources used and the calculation of the indicators are published.

The first phase includes the publication of 20 indicators in the second half of 2023, with annual information since 2010. In a second phase, new data sources and targets will be explored to add more indicators to the project.

4.3.3 The importance of energy statistics and the role of NSOs

315. The energy sector accounts for about three quarters of total GHG emissions globally, hence addressing the clean energy transition is central in achieving climate mitigation objectives. Accurate, timely and transparent energy statistics form the basis for energy and climate policy planning. Reliable data are now needed more than ever to track the progress of energy transitions.

316. While the international energy and climate community can provide financial and methodological support to the countries, the important work in collecting and developing energy statistics at the national level is the responsibility of the national administrations.

317. An important enabler of the development of energy statistics is the degree of political will and awareness and the priority given to the subject matter. The prioritization of energy statistics at a national level requires continuous awareness-raising among policy and decision makers about its importance and benefits. Once this recognition is established it becomes easier to justify proper resource allocation to the collection of data and the development of corresponding indicators.

318. Data collection corresponding to energy statistics should be grounded on a solid framework mandated and enforced by law. This allows there to be, simultaneously, one or more government institutions with clearly established responsibilities and tasks for data collection, and an obligation of businesses, industry and citizens (depending on the data collection target) to collaborate and provide the requested data in an accurate and timely manner, making data collection efforts feasible. A coordination mechanism can be established to ensure the efficient use of existing data and to avoid duplicity of work.

319. Primary responsibility for compiling official national energy information may fall under another entity than the NSO, such as the ministry of energy, the energy agency or similar bodies. In such cases, the role of NSOs may differ and the NSOs should be working on energy statistics in concerted arrangements with the other responsible bodies. In many cases, NSOs support energy ministries for instance by implementing dedicated energy surveys or integrating energy modules in existing surveys.

320. Energy statistics professionals often argue that the resources allocated to this topic are insufficient. It is important that governments allocate appropriate resources to energy data collection, and in particular demand-side energy consumption patterns, as these require additional detail and effort. The appropriate level of funding depends on a number of aspects related to the context. For example, the size of the country, existing data sources, how advanced data collection is and what the data gaps are, all represent important variables.

321. Thanks to their expertise on data collection, NSOs can play a pivotal role in primary energy data collection to inform sound decision-making required for achieving climate objectives both at national and international levels. Different approaches can be taken by NSOs to establish and maintain a sound national statistics system for energy. These approaches must be developed to fit within the local context and circumstances. However, regardless of the approach adopted, the national statistics system must be built upon a strong foundation that incorporates a solid data collection strategy, promotes both human and IT resources and benefits from effective institutional arrangements and international collaboration.

322. NSOs should adopt a well-designed energy data collection strategy which:

(a) Is planned in accordance with the national context and allocated budget

(b) Promotes dialogue between energy statistics and policy making to raise awareness of existing data needs and long-term objectives

(c) Facilitates institutional arrangements for data collection.

323. For NSOs it is important to ensure staff capacity and stability. Qualified staff who understand key energy concepts and the methodological particularities of energy statistics are an absolute requirement for the development of energy-related data at a national level. In addition, it is important to build capacity in different data collection methodologies (including administrative sources, surveys, modelling and metering), and how to derive indicators from the raw data collected.

324. The development of energy statistics at a national level is often a multi-stakeholder effort, as data are typically scattered and collected by different institutions. Given the scattered nature of energy data in particular in the case of end-use energy and activity data required to develop energy efficiency indicators, NSO should promote strong institutional collaboration and communication. In addition, organizational structures with a clear definition of responsibilities and clear communication channels can typically enable an easier and less costly data acquisition process. For example, existing data sharing arrangements and relationship building with counterparts in other institutions can facilitate the process. However, where there is significant bureaucracy to overcome for data sharing, institutional collaboration should be fostered at a high level. This arrangement will allow a higher level of engagement and accountability among stakeholders and can improve the consistency and efficiency of the collaboration.

325. Co-operation is also an important driver of improved data collection processes at an international level. This can be done, for instance, by inviting countries to share their experiences of data collection methodologies or the financing of data collection. It can also be done by joint work and common methodological frameworks for the development of indicators, in line with international methodologies, allowing for comparable data and findings.

Figure 4.22
From national energy data collection to energy-related climate indicators for policymaking

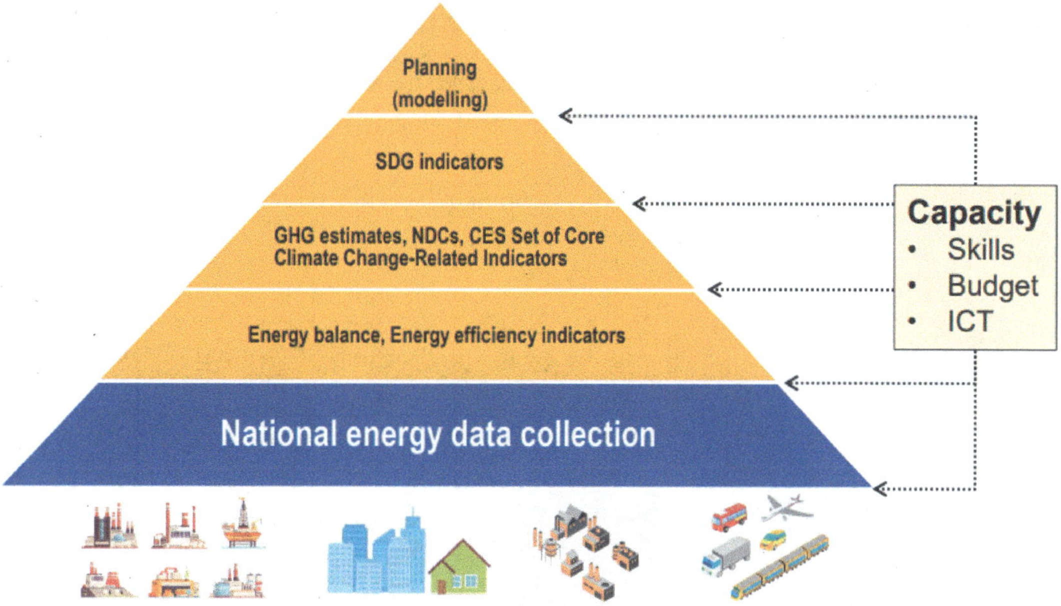

Source: IEA (2022a).

326. As depicted by Figure 4.22, primary energy data collection both on a supply and demand side are the perquisite for developing energy-related climate data and indicators which feed policy-making both at national and international levels. The energy balance is based on supply and demand data (United Nations, 2018). The energy balance is the basis for developing the GHG emissions associated with the energy sector. Additionally, merging economic activity data with energy balances allows developing energy efficiency indicators, indicators to track relevant Sustainable Development Goals (e.g. SDG 7.2.1, SDG 7.3.1 and SDG 9.4.1) as well as energy related NDCs.

327. Energy supply information can rather easily be obtained even on a monthly basis from tax registries (extraction), customs (trade) or other administrative sources set up to collect such data[67]. In contrast, collecting and compiling a full picture of the national energy demand potentially from thousands of enterprises/economic agents require steps of data validation, aggregation and preparations for dissemination.

328. The actual choice of the data collection method is specific to the national situation. A review of the data already available in the country (whether or not in the same institution or in other institutions in the country) is often a first step. If data are already available and can be used for energy statistics, there is little justification to embark on a new data collection. It is important however to have information on how the data are collected as it affects how fit for use, they are for statistical purposes. Classical examples include the use of administrative data from public or private registers, or data collected through an existing survey designed for different purposes. A typical example of the latter is data on diesel consumption for agricultural purposes that are collected through an existing agricultural statistics survey. Using the existing data helps to save resources and reduce response burden. If no data are available in the country from other sources and a new data collection method has to be put in place (whether adding questions to an existing survey or designing a new survey),

[67] For example, in Australia such data are available from the Australian National Greenhouse and Energy Reporting Scheme (NGERS): https://cer.gov.au/schemes/national-greenhouse-and-energy-reporting-scheme

additional work needs to be carried out to identify the respondents as well as the necessary human and financial resources.

329. Multiple types of data sources and data collection methods including administrative data, statistical surveys, modelling and measurements can be adopted by the NSOs as applicable. As summarized by Table 4.2, collection methods correspond to a set of advantages and disadvantages and should be carefully selected based on specific circumstances.

Table 4.2:
Summary of pros and cons corresponding to different data collection methods

Methodology	Pros	Cons
Administrative data sources	• Avoid cost of a new data collection process • Available relatively quickly • Increase synergies between institutions • Raise profile and interest of energy data among various services	• Boundary issues: potential mismatch between definitions and target populations of existing data and data needed • Challenges in establishing and maintaining communication with the source organization • Potential costs (direct and indirect: e.g. purchasing data, establishing agreements, adapting data formats) • One-off time investment in search for data sources
Surveying	• Relatively cost-effective, given extensive information collected • Ad-hoc design of items collected based on purpose • Representativeness/statistical significance • Overall, comprehensive and good quality information	• Potentially high absolute cost • Time consuming • Need for further estimation work (e.g. extrapolation between years) • Risk of incomplete responses, biases, sampling errors • Requirement of staff training
Measuring (metering)	• Provides actual energy consumption at end-use or equipment level • High accuracy of collected data • Can shed light on actual behavioural patterns • Can be a key complement to other methodologies	• High cost of equipment • Small sample of population and time/lack of representativeness • Possible malfunctioning of equipment • Difficulties in finding volunteers
Modelling	• Cost-effective • Designed based on purpose • Can consolidate data from multiple sources • Can provide estimates of variables that cannot be measured • Allows validation of bottom-up estimates against national energy statistics	• Relies on availability of input data • Depends on quality of input data • Depends on model assumptions • Transparency may be an issue

Source: Adapted from IEA (2014).

330. In practice, the data not available from existing survey and administrative sources can be collected through new or revamped surveys. In addition, emerging technologies such as smart consumption meters, GIS mapping and satellite imagery have the potential to complement surveying in the future. Ideally, data modelling (imputation) is only used to fill otherwise unavoidable data gaps to complete the full picture of the national energy supply and consumption.

331. Administrative microdata, such as utility meter data, energy performance certificates of buildings and vehicle test data are among new data sources which can complement traditional data collection methods. Such new sources of data increase the efficiency of data production and/or enable

producing new statistics and indicators. Such data can be particularly important for developing end-use data and efficiency indicators required for tracking progress towards energy efficiency measures as well as linking energy statistics with socio-economic information to develop indicators for tracking just energy transitions.

332. Energy efficiency is a central element of achieving carbon-neutral energy transitions. It also brings important co-benefits such as job creation, energy security, higher productivity, savings on energy bills, and improved comfort and wellbeing. Tracking efficiency progress is only possible with detailed data - at end-use level - and indicators that allow the disentangling of the effects of different energy consumption drivers (e.g. activity, structure and efficiency). The collection of end-use data and the development of energy efficiency indicators is not a straightforward process. It should be grounded on a robust and detailed data collection process and based on international methodologies in order to allow for comparability and interpretation across countries. The IEA Demand-side data and energy efficiency indicators (IEA, 2023) provides a guide to designing a national roadmap that can be used as a guidance for NSOs, describing options and good practices for the collection of energy end-use data and the development of energy efficiency indicators at the national level. This guide is a strategic document looking at the whole value chain in the development of efficiency indicators - from the initial point where the need for data and indicators arises, to the later dissemination and data use stages. It is intended to be a useful resource for practitioners across the world, assisting them in the process of developing a roadmap to energy efficiency indicators at the national level.

333. Without a major acceleration in clean energy innovation, climate mitigation targets are not achievable. Estimations show that reaching climate targets will require a widespread use of technologies that are still at the research, development and demonstration stage today. Major innovation efforts are vital so that the technologies necessary for net-zero emissions can reach markets as soon as possible. Tracking clean energy innovation progress encompasses several critical elements of effective energy innovation policy: identifying gaps and opportunities, evaluating the effectiveness of programmes and policies, and understanding the market readiness of key technologies. One of the policy tools available as part of national energy research policies and support programmes is direct public expenditures on RD&D. The only way to accurately track them is to have robust and reliable statistics. These data provide the quantitative basis to design, evaluate and adjust those policies. They also generate valuable insights for government ministries and departments that are not directly responsible for energy research spending. The IEA *Tracking Public Investment in Energy Technology Research: A Roadmap* (IEA, 2022b) has been produced to help NSOs design or revamp a robust data collection, validation, and dissemination process based on the experience of IEA member and association countries. This roadmap describes the variety of country approaches, also identifying the most important common components: the institutional arrangements; the methods for collecting, classifying and validating the data; the data management and technology processes; and, finally, the dissemination.

334. As discussed, the energy sector accounts for about three quarters of total GHG emissions globally, hence high-quality emission estimates corresponding to the energy sector are pivotal for monitoring progress towards national and global emission reduction targets. The energy balance based on the supply and demand data includes the majority of data required for developing the GHG emissions associated with the energy sector. According to the 2006 IPCC Guidelines, the main approach for estimating emissions from the energy sector is called the Sectoral approach. The scope of estimates based on the sectoral approach includes all GHGs. Additionally, a second methodology called the Reference approach can be followed as an independent check providing a quick and easy alternative estimate of only the CO_2 emissions.

335. According to the classification provided by the 2006 IPCC Guidelines, the energy related GHG emissions consists of three subcategories: i) emissions from fuel combustion, ii) fugitive emissions and iii) emissions from carbon dioxide transport, injection and storage. For estimating the fuel combustion emissions based on the Sectoral approach, granular demand-side data across the sectors are required which are typically available from the energy balance. The emission estimates may follow a simple Tier

1 methodology which is based on default emission factors provided in the 2006 IPCC Guidelines, or more sophisticated Tier 2 and Tier 3 methods which allow taking into account detailed country-specific or technology-specific information or direct measurements. A Tier 1 approach works relatively well to estimate CO_2 emissions, as carbon is typically conserved during fuel combustion. For estimating the fugitive emissions and emissions from CO_2 transport, injection and storage, additional activity data on top of energy balance are required.

336. The reference approach is a top-down approach based on the supply side of the energy balance. The approach is based on the principle of mass conservation, meaning that the carbon brought into a national economy in the form of a fuel is either released into the atmosphere as GHG, or is diverted (i.e., increased fuel stocks, stored in products, etc.). This means that the non-CO_2 GHG emissions from fuel combustion are excluded from the scope of this approach. Although official data require sectoral approach estimates, it is good practice to apply both a sectoral approach and a reference approach and to compare the results of these two independent estimates. Significant differences may indicate possible problems with activity data, net calorific values, carbon content, etc. For additional details on both approaches, please refer to the 2006 IPCC Guidelines for National GHG Inventories.

337. About three quarters of the GHG emissions globally are from the energy sector. The main sources of the remaining quarter are agriculture (CH_4 and N_2O), LULUCF (CO_2) and the fluorinated gases used in industry and transport (e.g., for refrigeration and air conditioning). To achieve the global and national mitigation targets (2030 and 2050) high quality statistics are necessary also for these domains, and NSOs should have a prominent role - similarly to energy statistics.

338. Not only a successful energy transition is essential to reach the mitigation objectives. Similar transitions are needed in, for example, transport and agricultural sectors. Less energy use is strongly encouraged in all such transitions, but in the agricultural sector many other necessary changes are not energy-related and focused on non-CO_2 emissions, e.g., the sustainable management related to livestock, fertilization and crop production. The transition of the transport sector also entails changes that are not energy-related, such as improved public transport, more international trains, and encouraging working from home (all leading to less road traffic, and less use of motor fuels).

4.4 Conclusions and recommendations

339. Section 4.2 presented an overview of (inter)national mitigation policies, including related data needs, whereas Section 4.3 outlined the types of data and statistical expertise NSOs can provide to policy makers and other data users.

340. The demand side (Section 4.2) included types and possible grouping of mitigation policies (Figure 4.1) and mitigation actions and options described in nationally determined contributions (Figure 4.2 and Figure 4.3) indicating need for data on Energy supply, Transport, Buildings, Industry, Agriculture, LULUCF, Waste, and Cross-cutting areas. These data needs are illustrated by country examples in Section 4.2.4. Complementary approaches to measuring GHG emissions (SEEA, see Figure 4.4 and Table 4.1) and the importance of carbon sequestration and storage by world's forests (REDD+; see Figure 4.5) are also mentioned.

341. The supply side (Section 4.3) offered a variety of statistics on GHG emissions and directly related topics at international level (Figure 4.6–Figure 4.15) illustrated by country examples (Figure 4.16–Figure 4.20). All examples are informative and well designed, but it should be noted that in general, at a global level, many of these examples are currently outside the competencies of NSOs.

342. The monitoring of mitigation policies should not only focus on reducing greenhouse emissions. Enablers that are essential for achieving successful transition(s) should also be monitored (Figure 4.21). In addition, the importance of energy statistics (and the role NSOs) needs to be emphasized, just like other statistics used as essential inputs for inventory compilations (as activity data) and other relevant statistical data to achieve successful transition(s).

343. It is overwhelming to see what is already available on both the demand side (Section 4.2) and the supply side (Section 4.3). It seems therefore too cumbersome to confront both sides to expose the data gaps at a granular level. Instead, a more high-level approach is used (Figure **4.23**). The figure shows that, depending on national circumstances and mandates, NSOs can have two main roles in the mitigation domain of the Enhanced Transparency Framework under the Paris Agreement: (1) to monitor GHG emissions, and (2) to monitor the mitigation policy actions in NDCs. In both roles, one of the challenges that NSOs face is the alignment of concepts, methods and classifications used under the Enhanced Transparency Framework with those used in official statistics.

344. The first NSO role is focused on collecting primary data that are relevant to climate indicators and inventory compilation, enabling providing activity data for the inventory compilation process. As part of this first NSO role, the statistical community enriched the IPCC emissions by adding complementary approaches, based on SEEA-CF, to further monitor the GHG emissions (residency versus territory approach; production versus consumption approach; indicators such as emission intensities and footprints). The second NSO role is less mature. NSOs can provide tailor-made data to track the progress of mitigation policy actions. However, more efforts, resources and capacity building are needed to close the gap between the lower level in Figure **4.22** (primary energy data) to the higher levels (energy-related climate indicators for policy making). For both purposes, energy statistics plays a central role.

Figure 4.23
Data from NSOs to monitor greenhouse gas emissions and mitigation policy actions

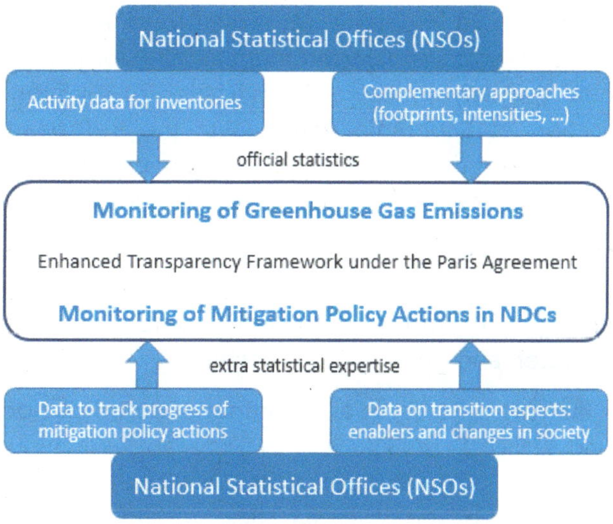

Source: Authors.

345. Similarly, more data is required on all kinds of transition aspects, not only for the clean energy transition but also for the transitions in other sectors (Transport, Agriculture, Buildings, and Industry). This concerns data requests about enablers that are essential to realize a successful transition. There is also a need for data on changes in the society as a result of transition processes, including the monitoring of the perception and behaviour of households, businesses and governments.

346. For both NSO roles, the compilation and provision of official statistics facilitates the monitoring of the mitigation domain in the Enhanced Transparency Framework under the Paris Agreement. To that end, NSOs can set up dashboards to disseminate harmonized indicators. Additional statistical expertise is indispensable in this respect, especially for the second NSO role, including investment in a stronger role in data stewardship, more use of administrative microdata (and data linking), and more collaboration with researchers, policy makers and other ETF stakeholders. For example, NSOs can act as independent data coordinators in national energy transition processes or actively participate in monitoring the government's implementation of the national net-zero plans.

347. One way to bridge the gap between both NSOs roles in Figure **4.23** and what is necessary in the ETF mitigation domain is to conduct detailed assessments, for example, by making use of the Climate Change Statistics and Indicators Self-Assessment Tool (CISAT) (UNSD, 2023). The combination of CISAT with the Global Set Implementation Guidelines (United Nation, 2024). support the countries in preparing their own sets of climate change statistics and indicators according to their individual concerns, priorities and resources. Such national exercise can be based on the overviews in Sections 4.2 (demand side) and 4.3 (supply side). This requires more dialogue and exchange between data users and producers at various levels, with ETF stakeholders in the mitigation domain having direct and regular contact with NSO representatives.

348. The main objective of such a dialogue between the ETF stakeholders and NSOs would be to bridge the data gap between what NSOs can contribute (supply side), and what is needed in the ETF mitigation domain (demand side), as depicted in Figure **4.23**. Making use of CISAT can be one of the tasks to undertake, although other tasks are probably more appropriate to start with. For example, the ETF mitigation policy data needs as presented in Section 4.2 require more detailed descriptions of what (micro)data and statistical expertise are really needed, including some additional guidelines on how the requested statistical data, as shown in section 4.3, can be produced and disseminated by NSOs (or other data producers).

349. To structure the first discussions in the 'ETF NSO Mitigation' dialogue the following strong points of NSO statistics compared to UNFCCC data can be taken into account:

(a) NSO statistics identify economic actors involved in emissions, which is something UNFCCC data does not do, i.e., record information on which businesses, in which economic activity, and the role of the households and government. NSO information is grouped by economic activity: International Standard Industrial Classification of All Economic Activities (ISIC) and Statistical Classification of Economic Activities in the European Community (NACE). A statistical approach based on businesses economic activity also allows to produce environmental footprints based on input-output modelling, e.g., to produce carbon footprints.

(b) GHG emissions can be related to official statistics directly linked to those emissions: energy use, energy production, renewable energies, transport statistics, electric vehicles, energy for heating, etc., including a link from economic activities to environmental impacts, like the use of materials, use of water, and generation of waste.

(c) Another aspect is the role of statistics about information on carbon sequestration. UNFCCC organizes it around LULUCF. NSOs can provide additional quantitative information on LULUCF elements, and to do so in a consistent way, such as forest (and biomass) statistics, forest accounts, land use and land cover statistics, and SEEA ecosystem accounts.

(d) There are lots of thematic statistics that are relevant here and can be given prominence: agriculture, forestry, transport, buildings, etc. Such statistics can cover topics like greenness of the international trade, offshore emissions, and international aviation & shipping.

(e) Statistics allow to link GHG emissions data with other economic aspects, notably about production and consumption. Statistics can complement UNFCCC with information about economic actors involved in the emissions, in particular whether they are businesses, households, or government. For businesses, it is possible to identify their economic activity, which in turn paves the way to complement with other economic variables about that economic activity, such as investments, jobs, exports, finance, taxes paid, and subsidies received. Statistics are well placed to cover those data, because they are directly based on business surveys, government finance statistics and national accounts.

(f) There are also differences in quality aspects, e.g. NSOs consider timeliness as one of the quality criteria (improves data usability), which is not the case in the inventory community (they have TACCC: transparency, accuracy, completeness, comparability, and consistency; which is also covered by the other NSO quality principles). E.g., SEEA quarterly estimates of GHG emissions based on proxies are available much sooner than official GHG inventory estimates, which use more detailed data.

350. This list of strong points of NSO statistics compared to UNFCCC data is not complete and can be extended by mentioning the strong points of UNFCCC data compared to NSO statistics, and to also include the strong points of other data producers.

351. To know each other's strengths and weaknesses helps to define the responsibilities of all the ETF stakeholders in the mitigation domain. An 'ETF NSO Mitigation' dialogue would reduce the risk of duplication of data production, facilitate identification and solving of data gaps, and also play an eminent role in selecting the 'best' dataset. Before setting up a new platform for dialogue, it is recommended to explore and assess the synergies with the processes already underway at national, regional and global levels.

352. Based on the conclusions as given above the following recommendations are made:

(a) More dialogue and exchange between data users and producers is needed to bridge the data gap between what NSOs can contribute (supply side) to what is needed in the ETF mitigation domain (demand side).

(b) Such 'ETF NSO Mitigation' dialogue can provide more detailed descriptions of ETF mitigation policy data needs, more guidance on producing the requested NSO data, and more effective collaboration in solving identified data gaps (e.g., by making use of CISAT).

(c) The NSOs need capacity building for further development and extensions of official statistics and other tailor-made (micro)data, in particular the variety of data needed to monitor the clean energy transition and other transitions envisaged in NDCs. Data stewardship by NSOs plays a central role here. This includes microdata linking and a consistent data compilation from different perspectives (production approach, consumption approach, and changes in society).

Chapter 5 Statistics for climate change adaptation policymaking

5.1 Introduction

353. The overarching purpose of this chapter is to identify what role NSOs can play in providing to policymakers data needed to make informed decisions on climate change adaptation and assess their effects. The statistical work on adaptation is far less mature than mitigation and more context specific, as the information needs depend on specificity of risks occurring in each country and the measures adopted to address them. For that reason, the chapter also aims to provide more clarity on what adaptation is and what data related to adaptation may encompass, and thus assist NSOs in better understanding and navigating data needs in their national circumstances.

354. The chapter starts with introducing key concepts related to climate change adaptation (section 5.2.1), the typical national adaptation planning process (5.2.2) and key challenges in implementing and assessing adaptation actions (5.2.3). Then, the chapter presents a list of key stakeholders and their potential role and provides an overview of adaptation data landscape. Against this backdrop, the second part of the chapter (section 5.3) discusses how NSOs can contribute to improving data for adaptation, focusing both on data that already is produced in established domains, as well as on new work that potentially could be undertaken. Section 5.4 draws conclusions and provide recommendations on the role of NSOs in this area.

5.2 Definitions, policy context and data landscape related to climate change adaptation

355. In the discussions about climate change policies, there is often a lot of focus on mitigation - limiting GHG emissions to slow down the global warming. However, as mentioned in section 2.1.1.1, the Paris Agreement also includes an equally important goal on climate change adaptation to

> *"strengthen the global response to the threat of climate change, in the context of sustainable development and efforts to eradicate poverty, including by (…) increasing the ability to adapt to the adverse impacts of climate change and foster climate resilience and low greenhouse gas emissions development, in a manner that does not threaten food production".*

356. As stated by the IPCC (2023a), human-caused climate change is already affecting many weather and climate extremes in every region across the globe. This has led to widespread adverse impacts and related losses and damages to nature and people, which will intensify with every increment of global warming. Action on adaptation is therefore just as important and urgent. Moreover, climate change mitigation and adaptation are inherently related. On one hand, integration of mitigation and adaptation strategies can offer important opportunities and trade-offs. On the other hand, insufficient progress with mitigation requires even bigger adaptation efforts.

357. What actually is climate change adaptation? According to IPCC (2022a)[68],:

> *"Adaptation is defined, in human systems, as the process of adjustment to actual or expected climate and its effects, in order to moderate harm or exploit beneficial opportunities. In natural systems, adaptation is the process of adjustment to actual climate and its effects".*

358. The potential for adverse consequences (risk) results from "interaction between climate-related hazards with the exposure and vulnerability of the affected human or ecological system". What constitutes effective adaptation depends heavily on what the risks and opportunities are in the given context. Informing and monitoring meaningful adaptation response requires therefore information about climate-related hazards, exposure to them, vulnerability and resilience. All these concepts and

[68] Unless indicated otherwise, the source of IPCC definitions is the Glossary of the IPCC Sixth Assessment Report, Climate Change 2022: Impacts, Adaptation and Vulnerability (IPCC, 2022a)

related data are relevant in this context and will be introduced in the section below. To be relevant and appropriate, decisions on adaptation response should thus be evidence-informed, taking into consideration context-specific situations and circumstances.

359. Climate change adaptation is linked with many socioeconomic issues and other policy areas. It is particularly closely related to sustainable development and disaster risk reduction. As laid out in Chapter 2, the 2030 Agenda, the Sendai Framework, and the Paris Agreement are interlinked. As described by IPCC:

> *"Adaptation plays a key role in reducing risks and vulnerability from climate change. Implementing adaptation and mitigation actions together with the SDGs helps to exploit synergies, reduce trade-offs and makes all three more effective. From a risk perspective, limiting atmospheric greenhouse gas (GHG) concentrations reduces climate-related hazards while adaptation and sustainable development reduce exposure and vulnerability to those hazards. Adaptation facilitates development, which is increasingly hindered by impacts and risks from climate change. Development facilitates adaptation by expanding the resources and capacity to reduce climate risks and vulnerability".*

5.2.1 Understanding climate change risks, impacts, adaptation and vulnerability

360. The sections below introduce the key concepts related to climate change adaptation.

Risk

361. As defined by the IPCC in the context of climate change, risk is:

> *"The potential for adverse consequences for human or ecological systems, recognizing the diversity of values and objectives associated with such systems. In the context of climate change, risks can arise from potential impacts of climate change as well as human responses to climate change. Relevant adverse consequences include those on lives, livelihoods, health and well-being, economic, social and cultural assets and investments, infrastructure, services (including ecosystem services), ecosystems and species.*
>
> *In the context of climate change impacts, risks result from dynamic interactions between climate-related hazards with the exposure and vulnerability of the affected human or ecological system to the hazards. Hazards, exposure and vulnerability may each be subject to uncertainty in terms of magnitude and likelihood of occurrence, and each may change over time and space due to socio-economic changes and human decision-making.*
>
> *In the context of climate change responses, risks result from the potential for such responses not achieving the intended objective(s), or from potential trade-offs with, or negative side-effects on, other societal objectives, such as the Sustainable Development Goals (SDGs). Risks can arise for example from uncertainty in the implementation, effectiveness or outcomes of climate policy, climate-related investments, technology development or adoption, and system transitions."*

362. This definition brings out four important points:

(a) Risk can arise from both potential impacts of climate change and human responses to climate change.[69]

(b) Risk related to potential climate change impacts results from an interaction between three variables - hazard, exposure and vulnerability, which may each change over time and space.

(c) Risk related to human responses to climate change results from these responses being not effective or their negative side effects on other societal objectives.

(d) The adverse consequences may affect in various ways individual people, society, economy and environment.

[69] The second type of risk is related to the transition risk.

363. To understand the concept of risk, it is important to understand its determinants: hazards, exposure and vulnerability. Figure 5.1 below illustrates the relationship between these concepts, as well as the risk and impacts.

Figure 5.1
Key determinants of climate related risks are hazards, exposure and vulnerability. Impacts are the consequences of realized risks.

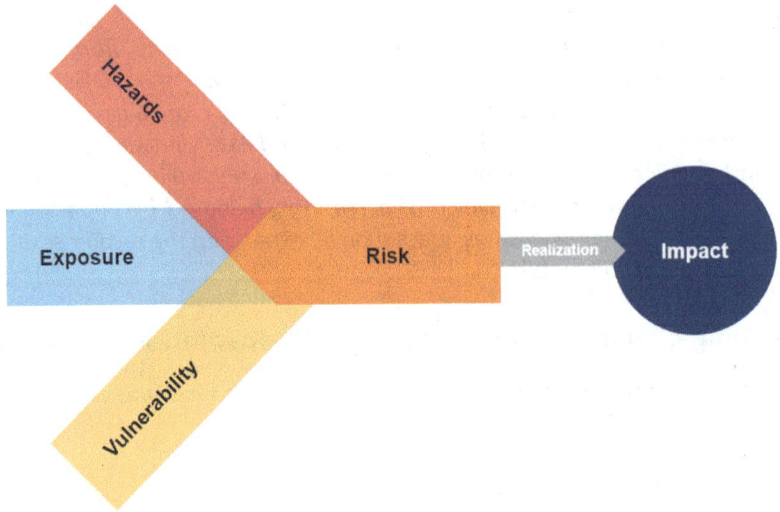

Source: Authors.

Hazard

364. Hazard is:

> *"The potential occurrence of a natural or human-induced physical event or trend that may cause loss of life, injury or other health impacts, as well as damage and loss to property, infrastructure, livelihoods, service provision, ecosystems and environmental resources." (IPCC, 2022a)*

365. Climate related risks are created by a range of hazards. Some are slow in their onset (such as changes in temperature and precipitation leading to droughts, or agricultural losses), while others happen more suddenly (such as tropical storms and floods).[70]

366. As part of the Nationally Determined Contributions (see Chapter 3 for more explanation on NDCs), many countries are communicating their adaptation priorities and actions. Most NDC adaptation components include key climate hazards faced by countries. UNFCCC analyzed, for the 2016 INDC Synthesis report, which hazards were mentioned in INDCs (Intended Nationally Determined Contributions - a term used before the adoption of the Paris Agreement). The results are shown in Figure 5.2. While such a quantitative analysis is not included in the latest NDC synthesis reports, countries continue to describe key climatic changes (in particular temperature increase, extreme temperatures, precipitation changes and sea level rise) which trigger extreme events (such as rainfall events, storms and cyclones), as well as flooding, drought, heatwaves, saltwater intrusion, ocean acidification, coral bleaching, erosion, landslides, fires and thawing ice and permafrost (UNFCCC, 2022a). Countries are undertaking hazard assessments to understand observed and projected hazards - as a way to inform their national adaptation plans (NAPs) (see section 5.2.2).

[70] UNFCCC website, Nairobi Work Programme https://unfccc.int/topics/resilience/resources/climate-related-risks-and-extreme-events)

Figure 5.2.
Key climate hazards identified in the adaptation component of the intended nationally determined contributions as of 2016 (number of Parties referring to a hazard)

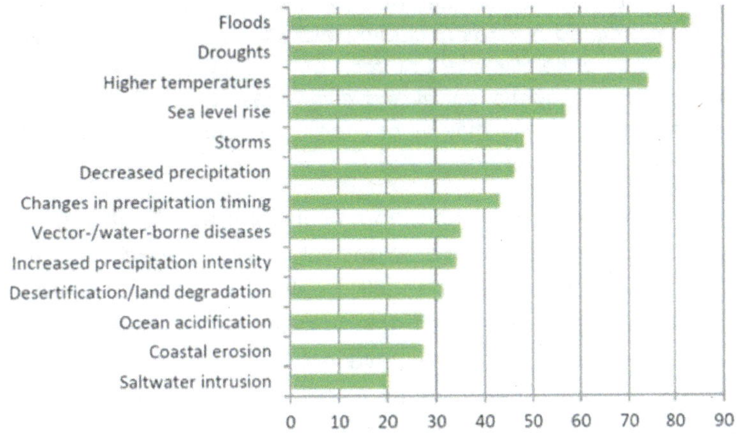

Source: UNFCCC (2016a).

Exposure

367. The second determinant of risk is exposure, which is:

> *"The presence of people; livelihoods; species or ecosystems; environmental functions, services, and resources; infrastructure; or economic, social, or cultural assets in places and settings that could be adversely affected." (IPCC, 2022a)*

368. Under a scenario with 1.5°C of warming above preindustrial levels by 2030, almost half of the world's population could be exposed to a climate hazard related to heat stress, drought, flood, or water stress in the next decade, up from 43 per cent today –and almost a quarter of the world's population would be exposed to severe hazards (McKinsey, 2021).

Vulnerability

369. The third determinant of risk is vulnerability, that is:

> *"The propensity or predisposition to be adversely affected. Vulnerability encompasses a variety of concepts and elements, including sensitivity or susceptibility to harm and lack of capacity to cope and adapt." (IPCC, 2022a)*

370. People who are socially, economically, culturally, politically, institutionally, or otherwise marginalized are especially vulnerable to climate change (IPCC, 2014). The rural and urban poor, and remote groups and communities have limited adaptive capacity due to their economic status, thereby making them more vulnerable than other members of the society. The young and the old are on average more vulnerable to climate change risks than other members of the population (UNFCCC Least Developed Countries Expert Group, 2018). It should be noted that the vulnerability may apply as well to ecosystems, infrastructure and other elements at risk.

371. The difference between exposure and vulnerability can be illustrated with examples. In the case of risks related to floods, people who live in an area that may be affected by a flood are exposed. In addition to exposure, a key risk determinant is also whether exposed people have characteristics that make them particularly susceptible to harm in the case of a flood. For example, poverty reduces people's capacity to adapt and cope if their accommodation is affected by the flood.

372. This reasoning also explains why GHG emissions increasing the likelihood and severity of hazards are not the only drivers of climate risk. Demographic and socio-economic developments, and ecosystem degradation, can also be climate risk drivers by increasing vulnerability or exposure.

373. The concept of vulnerability is also related to resilience and adaptive capacity (see the relevant section below).

Impacts

374. Impacts, in the context of climate change, are:

> *"The consequences of realized risks on natural and human systems, where risks result from the interactions of climate-related hazards (including extreme weather/climate events), exposure, and vulnerability. Impacts generally refer to effects on lives, livelihoods, health and well-being, ecosystems and species, economic, social and cultural assets, services (including ecosystem services) and infrastructure. Impacts may be referred to as consequences or outcomes and can be adverse or beneficial." (IPCC, 2022a)*

375. There is a strong conceptual link between risk and impact. An impact is a consequence of a realized risk, and a risk is the potential for adverse impact. As mentioned in both definitions, both risks and impacts can refer to effects on lives, livelihoods, health and well-being, ecosystems and species, economic, social and cultural assets, services (including ecosystem services) and infrastructure. For example, heatwaves and extreme temperatures have impacts on human health, infrastructure, such as transport networks, energy and health facilities and ecosystems (UNFCCC, 2022c). It is worth noting that this definition focuses on the impact of climate change and not on the impact of climate policies.

376. The impacts of climate change can be grouped in multiple ways, e.g. the IPCC groups them by the following seven sectors: Terrestrial and freshwater ecosystems; Ocean and coastal ecosystems; Water; Food, fiber and other ecosystem products; Cities, settlement and key infrastructure; Health, wellbeing and communities; and Poverty, livelihoods and sustainable development. These sectors can apply both for risks and impacts.

377. Another way of grouping could be by element at risk, as identified by the UNECE Task Force on measuring hazardous events and disasters: People, Housing, Basic services, Critical infrastructure, Economic activity, Ecosystems, Food security and agriculture, Water security, Energy security, Health care, and Cultural heritage (UNECE, 2023b).

378. Socioeconomic characteristics can help explain varying impacts of climate change experienced by various population groups due to differences in their vulnerability.

Resilience and adaptive capacity

379. The realization of risk leads to an impact, but its magnitude and scope will depend on the resilience and adaptive capacity of the "element at risk".

380. Resilience is:

> *"The capacity of interconnected social, economic and ecological systems to cope with a hazardous event, trend or disturbance, responding or reorganising in ways that maintain their essential function, identity and structure. Resilience is a positive attribute when it maintains capacity for adaptation, learning and/or transformation" (Arctic Council, 2016).*

381. Adaptive capacity is:

> *"The ability of systems, institutions, humans and other organisms to adjust to potential damage, to take advantage of opportunities or to respond to consequences" (Millenium Assessment, 2005).*

382. These two concepts are related and to some extent overlapping, and also related to vulnerability. Building climate resilience involves all actors (governments, communities and businesses) having the capacity to anticipate climate risks and hazards, absorb shocks and stresses, and reshape and transform development pathways in the longer term (UNFCCC, 2021a).

Box 5.1
Terminological differences between disaster risk reduction and climate change domains.

This report uses terms as defined by IPCC and UNFCCC as the leading United Nations bodies for climate change science and climate change policy respectively. The work of United Nations Office on Disaster Risk Reduction (UNDRR), the leading UN body on disaster risk reduction, is related to the work on climate change. The terminology and definitions established in the context of the UNDRR work (United Nations General Assembly, 2016), also referenced in the UNECE work on statistics on disaster risk reduction, are, however ,not always consistent with the terminology established by IPCC and UNFCCC. It is important to be aware of these differences when communicating with (statistical) experts, across these related fields. Three examples are presented below.

1. Risk and disaster risk

UNDRR defines disaster risk as the potential loss of life, injury, or destroyed or damaged assets which could occur to a system, society or a community in a specific period of time, determined probabilistically as a function of hazard, exposure, vulnerability and capacity. While the concept of risk used in the context of disaster risk and climate change seem similar, there are important differences. First, disaster risk has an additional determinant - capacity, which is defined as the combination of all the strengths, attributes and resources available within an organization, community or society to manage and reduce disaster risks and strengthen resilience. These characteristics are considered part of vulnerability as defined by IPCC, as vulnerability encompasses a variety of concepts and elements, including lack of capacity to cope and adapt. Second, risk, as defined by IPCC, includes both risk from climate change impacts and risks from climate change responses. The latter is not included in the UNDRR disaster risk definition.

2. Mitigation

The term "mitigation" is defined by UNDRR as the lessening or minimizing of the adverse impacts of a hazardous event. In the context of climate change, this term is understood much more narrowly, as a human intervention to reduce emissions or enhance the sinks of greenhouse gases. Actions "to mitigate disaster risk" would therefore include actions that would fall under both climate change mitigation and adaptation as defined by IPCC.

3. Resilience

Finally, in the disaster risk reduction context, resilience of a system is its ability to resist, absorb, accommodate, adapt to, transform and recover from the effects of a hazard in a timely and efficient manner. In the climate change context it is "the capacity of interconnected social, economic and ecological systems to cope with a hazardous event, trend or disturbance, responding or reorganising in ways that maintain their essential function, identity and structure".

Adaptation

383. As already mentioned in the introduction to this section, adaptation is

> *"In human systems, the process of adjustment to actual or expected climate and its effects, in order to moderate harm or exploit beneficial opportunities. In natural systems, the process of adjustment to actual climate and its effects; human intervention may facilitate adjustment to expected climate and its effects." (IPCC)*

384. Adaptation refers to changes in processes, practices and structures to moderate potential damages or to benefit from opportunities associated with climate change. In simple terms, countries and communities need to develop adaptation solutions and implement actions to respond to current and future climate change impacts.[71]

> *"Adaptation plays a key role in reducing exposure and vulnerability to climate change. […] In human systems, adaptation can be anticipatory or reactive, as well as incremental and/ or transformational. The latter changes the fundamental attributes of a social-ecological system in anticipation of climate change and its impacts. Adaptation is subject to hard and soft limits." (IPCC)*

[71] See https://unfccc.int/topics/adaptation-and-resilience/the-big-picture/introduction

385. Adaptation action can reduce climate change related risk, by reducing the hazard, vulnerability or exposure, or increase the adaptative capacity and resilience. It is worth noting that adaptation actions alone are not enough to avoid all the risks (i.e., there are limits to adaptation). Mitigation action also decrease climate change related risk by reducing the drivers of climate hazards - the GHG emissions - and thus the likelihood and severity of hazards.

386. The risk that remains following adaptation and mitigation efforts is referred to by IPCC as residual risk. Figure 5.3 illustrates how mitigation and adaptation actions aim to reduce climate risks and impacts.

Figure 5.3.
How adaptation and mitigation action can reduce climate related risks and impacts

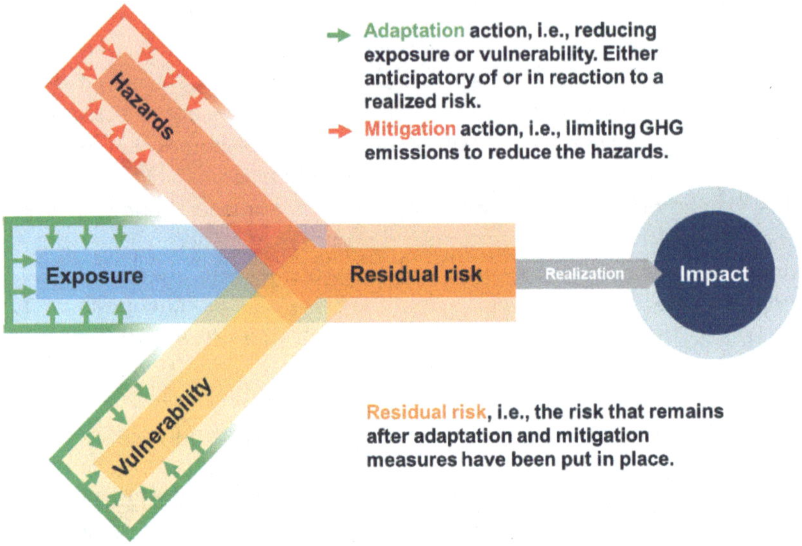

Source: Authors.

387. Climate change can also be a risk amplifier when combined with other potential natural disasters, or human driven actions. Consequently, in some jurisdictions, adaptation interventions require a cross-cutting understanding of these risk interactions and amplifiers.

388. It is important to highlight, that the definitions of impact and adaptation recognize the possibility of beneficial outcomes of climate change. Actions that enable benefitting from the opportunities associated with climate change are also part of adaptation, e.g. changing to crop varieties that are more suitable for the changed climate. The positive consequences of climate change are not presented on Figure 5.3 above, which focuses on the risk.

Box 5.2.
Adaptation actions in nationally determined contributions

Adaptation actions can have many types, e.g. technological/infrastructural, institutional, behavioural/cultural, nature based. Countries are undertaking or are aiming to undertake a variety of adaptation actions across adaptation priority areas and sectors.

In its 2022 NDC Synthesis Report, UNFCCC summarizes the priority areas and adaptation measures indicated by the Parties. For example, to reduce risks related to freshwater resources (identified by 87 per cent of parties as a priority area), countries are planning to strengthen watershed management, efficiency of water use and irrigation, via integrated water resource management, protecting and restoring water-related ecosystems such as forests wetlands and rivers, and diversifying water supply.

Most NDC adaptation components (81 per cent), mention the protection of terrestrial ecosystems and forests as a priority area. Planned adaptation actions are, for example, increasing protected areas, restore degraded lands and ecosystems and implement forest management and reforestation. Parties are also aiming to increase the resilience of infrastructure (55 per cent) through risk assessment, building codes and resilience standards, elevating infrastructure, associated education and nature-based solutions.

Figure 5.4
Share of adaptation components of nationally determined contributions referring to specific adaptation priority areas and sectors.

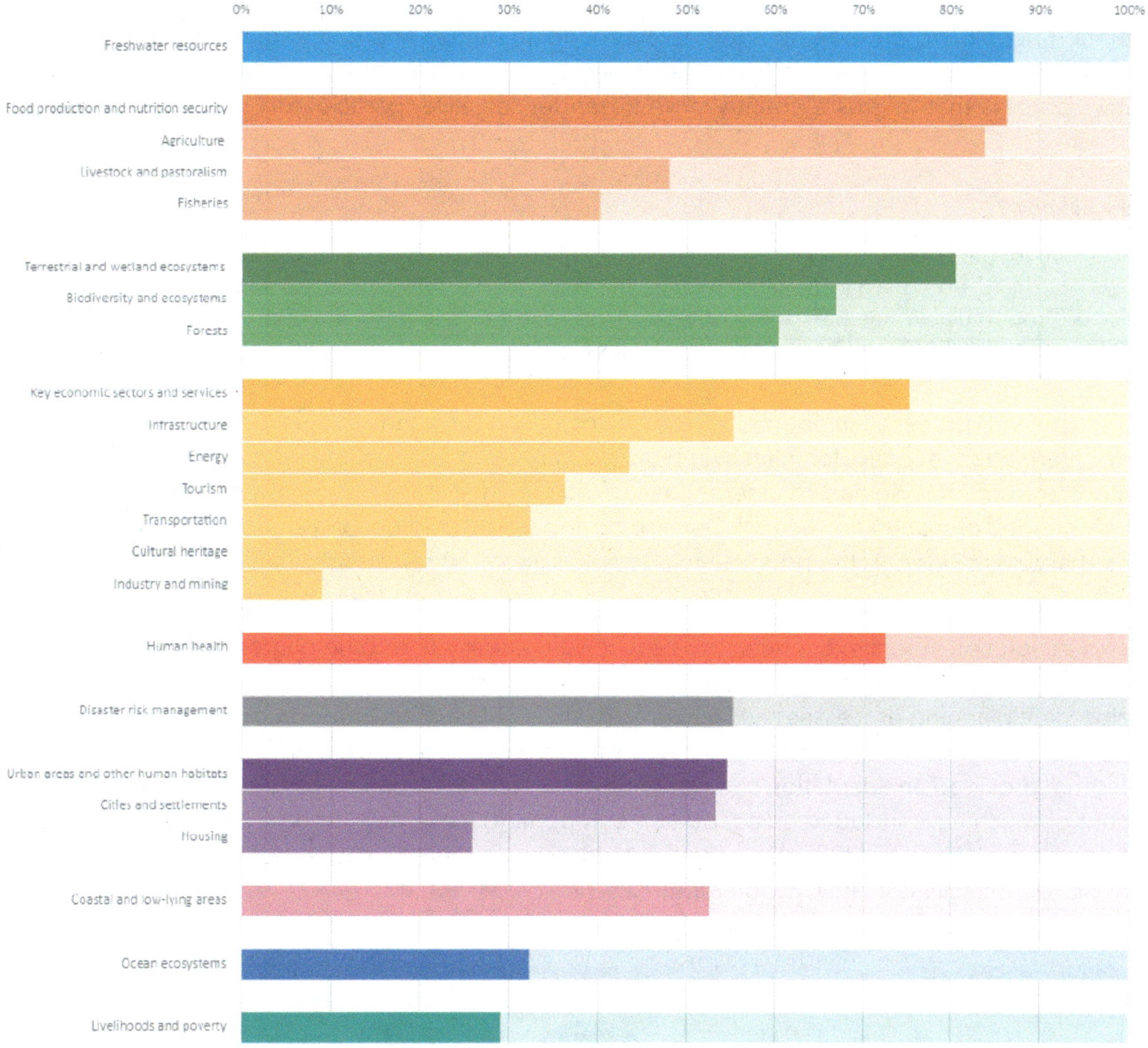

Source: UNFCCC (2022a).

Successful adaptation versus maladaptation

389. As mentioned in the section about risk, human response to climate change can also be a source of risk. Both adaptation and mitigation actions may not achieve their intended results or have unintended side-effects on/trade-offs with other societal objectives. Considering this aspect helps to distinguish between successful adaptation and maladaptation.

> *"Maladaptation refers to potentially adverse effects of certain forms of adaptation action, such as increased GHG emissions or increased vulnerability to climate change and diminished welfare of certain parts of a population now or in the future" (IPCC).*

390. As explained above, the Sustainable Development Goals and the Paris Agreement are interlinked. It is important to recognize, that actions aimed at sustainable development may undermine adaptation, and adaptation may have negative impacts on sustainable development, but both can be synergistic as well.

5.2.2 National adaptation planning and implementation process

391. The NAP process was established under the Cancun Adaptation Framework.[72] It enables Parties to formulate and implement NAPs as a means of identifying medium- and long-term adaptation needs and developing and implementing strategies and programmes to address those needs (UNFCCC Least Developed Countries Expert Group, 2012). It is a continuous, progressive and iterative process that follows a country-driven, gender-sensitive, participatory and fully transparent approach.[73]

Did you know?

A data plan for the NAP M&E should be developed and supported with adequate resources and data should be collected and stored in a systematic manner. The responsibility for compilation and aggregation of data collected from multiple agencies should be clearly assigned to one agency, which could be the national statistical office (UNFCCC Least Developed Countries Expert Group, 2012).

392. The NAP process can, for this report, be considered as a useful tool to understand the scope of the information needed for effective climate change adaptation policy making. For example, countries are undertaking hazard assessments on observed and projected hazards, and statistics are not only needed on adaptation actions, but also on impact, risk and its components. The main purpose of this chapter is to identify the potential role of NSOs in providing this information.

393. NAPs can enable countries to assess their vulnerabilities, mainstream climate change risks and address adaptation. The aim of NAPs is to reduce vulnerability and to help integrate climate change adaptation into relevant new and existing policies, programmes and activities in particular development planning processes and strategies (UNFCCC Least Developed Countries Expert Group, 2012). NAPs are a very important and increasingly prevalent policy tool. In 2022, 23 per cent of all parties that provided an adaptation component in their NDC (166 parties have submitted an NDC, 80 per cent of which have included an adaptation component), have developed a NAP, and 45 per cent intend to do so UNFCCC (2022a). Figure **5.5** below shows the adaptation policy cycle and support offered under the UN Climate Change regime.[74]

Figure 5.5
Adaptation Policy Cycle

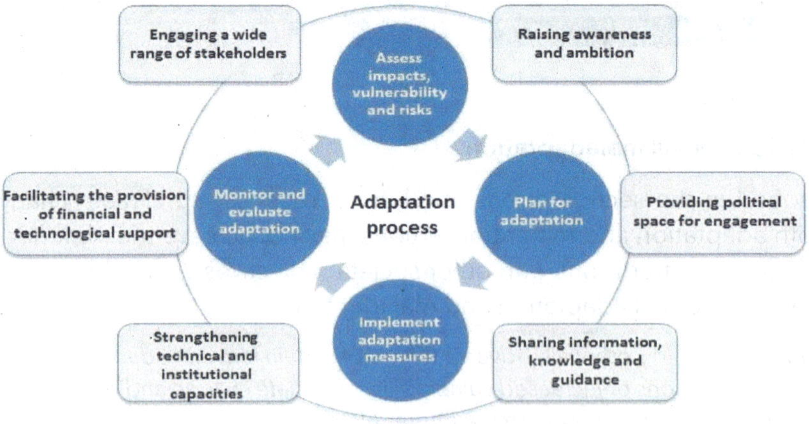

Source: UNFCCC[75].

[72] https://unfccc.int/process/conferences/pastconferences/cancun-climate-change-conference-november-2010/statements-and-resources/Agreements
[73] https://unfccc.int/topics/adaptation-and-resilience/workstreams/national-adaptation-plans
[74] More information on the adaptation cycle if required can be found in UNFCCC Adaptation Committee (2019).
[75] https://unfccc.int/topics/adaptation-and-resilience/the-big-picture/introduction

394. **Assessing impacts, vulnerability, risks, and resilience.** This first step consists of initial and periodically updated assessments, grounded in systematic observations and the best available science, of how climate change is impacting, and is expected to impact, natural and human systems. It also includes an assessment of the capacity of these systems to adapt to climate change impacts.

395. **Planning for adaptation**. Subsequently, it is necessary to identify potential adaptation measures, and evaluate these measures - using, for example, a cost-effectiveness analysis - to prioritize and choose among the options available. Adaptation planning should seek to avoid duplication, prevent maladaptation, and contribute to sustainable development outcomes. This step, along with the following implementation step, is undertaken while acknowledging that, in some cases, not all impacts will be addressed by planned adaptation, the adaptation measures could themselves lead to other negative consequences and there will be a role for contingency arrangements and loss acceptance.

396. **Implementing adaptation measures**. Once a plan is in place, the next step is to implement adaptation initiatives at various levels - including local, national and regional - and through a variety of means, including projects, programmes, policies or strategies. Increasingly, adaptation is pursued through mainstreaming initiatives into budgets, sectoral policies, and sustainable development plans, though efforts can nonetheless be undertaken as stand-alone processes.

397. **Monitoring and evaluating (M&E) adaptation.** Finally, a continuous process of monitoring and evaluating adaptation is undertaken. This helps ensure that efforts are successful and that the knowledge and information generated from planning and implementation are fed back into the process to bolster the success of future efforts. Monitoring seeks to keep a record of progress made throughout the implementation phase, while evaluation seeks to determine the effectiveness of the initiative.

398. Throughout this process, continuous stakeholder engagement and communication, along with capacity building, finance and technology, are essential to the success of each step.

For example

The European Union member States report on national adaptation actions every 2 years under the 2018 Regulation on the Governance of the Energy Union and Climate Action. The reporting is presented in Climate-ADAPT, the European platform on adaptation (developed by EEA on behalf of the European Commission) under the Country profiles (updated March 2023).

5.2.3 Challenges in implementing and assessing adaptation

The key types of challenges, gaps and barriers related to adaptation mentioned by the countries involve institutional arrangements, knowledge and capacity, data and information, policy, and resources (UNFCCC, 2022c).

399. Table 5.1 provides some examples of the challenges in each category presented in the report.

400. Although the challenges related to data and information are the most directly relevant to this chapter, worth noting are also mentioned challenges related to insufficient coordination among sectors and levels of government, public awareness (see role of NSOs in informing the public in Chapter 7), and resources for monitoring.

Table 5.1
Challenges, barriers and gaps linked to adaptation, as described in Parties' national reports

Type of challenge, barrier or gap	Examples
Institutional	Lack of or weak institutional infrastructure and frameworks
	Insufficient coordination of sectors and/or levels of government
	High staff turnover
Knowledge and capacity	Insufficient sector-specific training
	Gaps in knowledge on climate change and adaptation
	Limited knowledge-sharing among stakeholders
	Insufficient public awareness
	Insufficient technological capacity and knowledge
	Lack of specialized skills for advanced or scalable technologies
	Lack of local government capacity to implement adaptation
	Gaps in methodologies related to applying specific adaptation measures
	Insufficient uptake and application of traditional knowledge
Data and information	Lack of long-term reliable and up-to-date data
	Insufficient data on specific areas, such as climate impacts on health and well-being through a gender perspective, or climate induced migration
	Inadequate systems for data collection, management, reporting and verification
	Lack of research on applicable policy measures
	Limited access to, or insufficient quality and quantity of, necessary or basic adaptation-related information
	Poor dissemination and application of scientific findings related to climate change impacts
Policy	Weak mainstreaming of adaptation in national and sectoral plans and policies
	Lack of related legislation
	Insufficient addressing of climate change impacts in sectoral policies and legislation
Resources	Lack of human and financial resources for research and monitoring
	Lack of long-term predictable and adequate financing

Source: (UNFCCC, 2022c)

401. The UNFCCC Adaptation Committee also identifies data for adaptation as one of the technical areas relevant for the adaptation process (see the section on NAPs) and summarizes lessons learned and good practices, challenges, gaps and opportunities (UNFCCC Adaptation Committee, 2022). The following gaps and challenges are specifically identified:

(a) Gaps in observational and historical climate data in some regions of the world;

(b) Lack of sufficiently granular data from global climate models, suitable for adaptation planning to local level or for multisectoral, multipurpose measures, e.g. in coastal areas;

(c) Gaps in socioeconomic data, especially collocated with climate observations (see also box below);

(d) Challenges related to trade-offs between increasing timeliness, granularity and availability of adaptation data whole complying with quality standards.

Did you know?

UNFCCC Adaptation Committee (2022) notes that gaps in socioeconomic data, lack of international arrangements for coordinating collection of socioeconomic data under the climate regime and an absence of collocated time series of climate observations and socioeconomic parameters could in some cases inhibit a more accurate identification of climate risks and the attribution of observed changes to either climate impacts or socioeconomic developments.

Enhancing open access to existing climate and socioeconomic data, exploiting the potential of, among others, national statistical systems, and establishing international arrangements for the coordination of socioeconomic data collection and management, e.g. new adaptation-relevant official statistics are considered potential opportunities for enhanced action, support and international cooperation.

402. One very specific area which poses significant challenges in achieving adaptation objectives is assessing the progress in adaptation. The issues include lack of consensus on definitions and approaches surrounding adequacy and effectiveness, and significant gaps in monitoring and evaluation. Unlike certain other facets of climate change such as temperature rise or GHG emissions, adaptation poses a challenge of not being inherently quantifiable. Providing an accurate answer to a question whether a country is resilient to climate change is very challenging for a number of reasons.

403. Craft and Fischer (2018) identified some key challenges when reviewing progress on adaptation. Specifically, designing a system that can aggregate results across scales and contexts, satisfying the dual mandate of Global Stocktake of assessing collective progress and informing the update and enhancement of national level actions, overcoming the methodological challenges inherent in evaluating adaptation, such as the difficulty of attributing results to interventions and the shifting baselines and uncertainties of climate hazards, and navigating divergent views and political sensitives surrounding measurement under the UNFCCC regime. Tompkins et al. (2018) also hold that methodological challenges are one key problem area for assessing adaptation under the global stocktake but argue that empirical challenges (the rarity of adaptation databases) and conceptual challenges (lack of agreement on what counts as adaptation) are two additional core difficulties that must be considered.

404. Dilling et al. (2019) also warn of three challenges that stand in the way of assessing adaptation progress when it is framed as assessing the "success" of adaptation measures. First, the authors note, is the absence of a single, common definition of adaptation success or effectiveness.[76] Second, different perceptions of what constitutes successful adaptation may arise from different, and evolving, perceptions of what constitutes a risk and varying risk tolerances. Third is the challenge of existing power asymmetries and how they influence who can define what is measured when assessing adaptation success. Similarly, Leiter and Pringle (2018) caution that value judgments necessarily enter into the equation when assessing the success of adaptation action because it is possible that cases arise where, for example, adaptation undertaken by one demographic undermines the capacity of another demographic to adapt.[77]

[76] Absence of a single definition of adaptation might not be a negative element as adaptation is highly context specific. Any single indicator will not provide a complete picture of progress and using more than one tool to assess progress can provide a fuller picture of change.

[77] It is also important that potential indicators are assessed to ensure they don't lead to unintended consequences, such as supporting short-term action that might be detrimental to longer-term objectives, or that might lead to maladaptive practices or lock-ins. This is particularly important when looking at indicators developed in a non-climate context.

Did you know?

At COP28, in 2023, the UAE Framework for global climate resilience was adopted, marking a landmark milestone for adaptation under the UNFCCC and Paris Agreement process, which will guide both global and national adaptation moving forward. The framework includes qualitative thematic targets around water, food, health, ecosystems, infrastructure, poverty and cultural heritage, as well as quantitative and time bound targets around the adaptation cycle, as outlined in figure 5.5 above (assessment, planning, implementation, MEL).

The COP28 decision also includes guidance on reporting for the targets, inviting Parties to voluntarily include in their adaptation communications, biennial transparency reports, national adaptation plans, national communications and nationally determined contributions quantitative and/or qualitative information related to the targets. In many cases, NSOs will play an active role in this reporting process.

Indicators for the dimensional targets can be relatively straightforward (e.g. has a country conducted up-to-date assessments of climate hazards?), however the process for setting indicators for the thematic targets will be further elaborated in a 2-year UAE - Belém work programme (2023 to 2025). This will include identifying where existing data and indicators such as under the SDGs or Sendai Framework can be utilized, as well as identifying gaps and developing new indicators where required.

405. Most research and assessments related to climate change risk and adaptation have focused on the local-to-national dimension without accounting for the climate change impacts that cross national borders through, for example, trade, finance, people, and biophysical pathways such as changes in transboundary ecosystems. Consideration of transboundary risks and efforts could help capture a holistic picture of adaptation progress and avoid creating blind spots in cases where climate risks are redistributed rather than reduced.

5.2.4 Non-exhaustive list of stakeholders and their potential role

406. This section aims to highlight the main stakeholders holding climate data and the role they can play in helping the societies to adapt to climate change. However, the final list of stakeholders for a given country will depend on the country's context and public sector set-up. Many of the listed stakeholders might also be relevant to different areas of climate change. Therefore, identifying their roles in adaptation can help NSOs and other actors differentiate how the same stakeholders might play different roles in other areas relevant to climate change, such as mitigation.

407. When listing stakeholders of climate change-related data in adaptation or other relevant areas to climate change, prioritizing a set of key stakeholders is crucial particularly in low-capacity contexts. This prioritization could be based on criteria such as the stakeholder's capacity to identify relevant and already available information, produce and share specific data a country is interested in and the relevance of such data in the decision-making and reporting processes. Stakeholders can be selected depending on their capacities, available data, whether it the data is fit for purpose and other aspects a given country considers appropriate.

408. Relevant stakeholders are clustered according to the prospective role they can play around climate change adaptation data. In many cases, roles can overlap. Recognizing these differences can help NSOs better understand who to include in the collection, production, dissemination, and use of climate adaptation data.

Data providers and producers

409. Primary data providers and producers to be considered by NSOs are within the boundaries of a country's national statistical system. For example, environment or climate change units within the NSS may already possess information and data critical for adaptation. Beyond the NSO, principal data providers and producers might include actors from the following agencies[78]:

[78] The list is built based on the data providers highlighted in the Adaptation area of the Global Set of Climate Change Statistics and Indicators.

410. Line ministries:

(a) **Ministry of environment or ministry of natural resources** - information on monitoring and reporting on the national adaptation actions and other critical areas for adaptation, such as ecosystems and biodiversity.

(b) **Ministry of finance** - budget and resources dedicated to adaptation measures are useful indicators of the level of implementation and commitment. The Ministry of Finance might be able to provide government expenditure data on adaptation and environmental protection.

(c) **Ministry of labour and social affairs**- data relevant for adaptation to climate change, including impacts on vulnerable communities and children and women's representation and participation in adaptation sector-specific environmental governance bodies.

(d) **Ministry of education** - data to capture the extent to which education for sustainable development, including adaptation measures, could be provided by the Ministry of Education.

(e) **Ministry of public works or ministry of infrastructure** - information on public infrastructure projects, green urban areas and other adaptation measures being implemented in urban spaces.

(f) **Ministry of agriculture** - information on the agricultural sector's overall adaptation measures and sustainability. For example, how the sector responds and adapts to increased droughts, floods, or rainfall variability. .

(g) **Ministry of interior/internal affairs and/or ministry of defence** - depending on the country one of these ministries may usually be responsible for crisis management, disaster risk response, and protection of civilians during a crisis/emergency

(h) **Ministry of tourism** - data on the adaptive capacity of the major tourism sub-sectors, data on Infrastructure Resilience from tourist facilities and accommodations, or green infrastructure

411. Government agencies, offices and departments:

(a) **Disaster management agency or office/department of civil defence** - data on risk management process and risk reduction measures and strategies applied in the country.

(b) **Meteorological office** - data on temperature, wind, clouds, air quality and weather crucial for planning, implementation, and expansion of climate adaptation measures.

(c) **National geospatial authorities** - space-technology-based data, in situ monitoring, and reliable geospatial information.

(d) **Forestry department** - information on forest management covers crucial aspects for climate adaptation. For example, the forestry department might be able to provide data on the forest's level of sustainability regarding the administrative, legal, economic, social and scientific aspects that ensure the forest supplies goods and services relevant for present and future needs.

(e) **Migration agency** - data on population movement due to climate change (could also be under the ministry of interior).

(f) **Maritime authority and fisheries department** - data on coastal and marine zones and fisheries management.

(g) **Central banks** - data related to property and businesses at risk, and physical and transition risk.

(h) **National insurance regulatory authorities** - insurance can provide incentives to reduce risk and drive sustained adaptation measures. Hence, data on insurance premiums incurred due to climate-related events have a relevant role in the adaptation process.

(i) **Maritime authority and/or ministry of water/national water agency or similar body** – data on water quality and general characteristics of the ocean, freshwater resources and wastewater.

(j) **Waste authority and local government authorities** – either both or one of these stakeholders may be able to provide data on municipal waste collection and management.

412. Beyond the NSS[79] there are other actors such as non-government organizations, the private sector, development partners, communities or individuals that could provide data, contribute to the collection of data by NSS actors, or play other critical roles in advocacy or resource mobilization in aspects relevant for adaptation.

413. **Climate services** provide crucial climate information that assists individuals and organizations in society to make improved ex-ante decision-making, at national or international level. For example, Copernicus Services provide data on hazards and impacts on global scale.

414. Identifying actors beyond the NSS might be challenging for NSOs as, most of the time, no clear roles are defined for such actors. Hence, as proposed by the PARIS21 (2024), a starting point can be through the lenses of the following potential roles these actors can play in supporting the production and user climate change data.

(a) **Data users -** denoting data consumers which have the potential to use, process and analyse climate change data. These be at a global, national, or local level and comprise policy makers and governments, academia, think tanks and research institutes, private sector, civil society, non-governmental organizations, media, infomediaries and intermediaries, etc.

(b) **Other potential data providers/producers beyond the NSS** - referring to non-government organizations and those considered to be outside of a NSS that generate and store climate change data relevant to the country. These organizations might include civil society, academia and research institutes, private sector, infomediaries and intermediaries, civil society through citizen-generated data, etc.

(c) **Partners for developing skills and knowledge** - potential partners with strong technical expertise on climate change that can collaborate with the NSO and its partners to develop specific skills and valuable knowledge for any process along the data value chain. This might include, for example, data collection processes in a particular area related to climate change, data storage, management, the use of novel data sources to complement existing statistics, etc.

(d) **Advocacy actors** – encompassing groups and organizations involved in the processes of sharing, disseminating and communicating climate change data for raising awareness. These may include civil society, NGOs, media, infomediaries and intermediaries etc.

(e) **Resource mobilization actors** - organizations and entities providing funding and technical assistance for capacity development linked to climate change data and statistics production, use and dissemination, including development partners, international organizations, government agencies, donors, etc.

415. Particularly in jurisdictions with indigenous populations, there is a need for specific consideration to be given to the rights of these populations, the interactions with the wider governance system in the given jurisdiction, and the unique needs of these populations, addressed in culturally and legislatively appropriate ways.

[79] National statistical systems (NSS) are understood as the combination of statistical organisations and units within a country that jointly collect, process and disseminate official statistics on behalf of a national government. https://paris21.org/national-statistical-system-nss, see also section 2.2.1.

5.2.5 Adaptation data landscape

416. Adapting to climate change consequences protects people, homes, businesses, livelihoods, infrastructure and natural ecosystems. Developing, implementing and monitoring meaningful adaptation measures requires data and information about climate-related hazards, exposure to them, vulnerability and resilience.

417. Many countries are making explicit plans on enhancing data, information and monitoring, to enable research relevant for adaptation and ensure that adaptation decisions are informed by science. In their NDCs, parties are, for example, mentioning plans to develop data collection programmes, undertake a national census on impacts, developing integrated climate information systems, risk maps with climate data and scenarios, and data pooling and sharing platforms for accessing information (UNFCCC, 2022a).

418. Adaptation solutions take many shapes and forms, depending on the unique context of a community, business, organization, country or region. There is no 'one-size-fits-all-solution'– adaptation can range from building flood defences, setting up early warning systems for cyclones and switching to drought-resistant crops, to redesigning communication systems, business operations and government policies[80].

419. As shown in Figure 5.3 in the previous section, adaptation aims to reduce vulnerability or exposure or increase adaptive capacity and resilience, and through that, to reduce risk and adverse impacts of climate change and climate change responses. Consequently, the data relevant for adaptation also include statistics and indicators on hazardous events and exposure to them, vulnerability and impacts. Moreover, it should also be kept in mind that statistics relevant for adaptation can also be relevant for other areas, e.g. forest statistics can describe climate change drivers, impacts and adaptation actions.

420. As adaptation solutions evolve so will do the demand for data and statistics specific to adaptation and its use. For example, tools such as the Climate-ADAPT database[81] allow us to have a general view of how European countries across different sectors are implementing adaptation practices under various conditions. Based on the most popular practices, demand for data on biodiversity, water management and disaster risk reduction arise as a trend. Another emerging area is health.

421. The UNSD Global Set of Climate Change Statistics and Indicators identifies some of the most common statistics and indicators related to vulnerability, adaptation and impacts, which can be found in Annex 4 and on the Global Set web pages. The topics covered under each of these areas illustrate the breadth of the adaptation data landscape (Table 5.2).

[80] https://unfccc.int/topics/adaptation-and-resilience/the-big-picture/what-do-adaptation-to-climate-change-and-climate-resilience-mean

[81] https://climate-adapt.eea.europa.eu/en/knowledge/tools/case-study-explorer

For example

At the EU level, Climate-ADAPT summarizes tools, products and data useful for different audiences:

- The European Climate Data Explorer gives access to a high number of hazards and impact at European, national, subnational and transnational levels (in collaboration with Copernicus).
- The European Climate and Health Observatory provides access to a wide range of relevant publications, tools, websites and other resources related to climate change and health.
- Economic losses and fatalities: This dashboard considers the total and insured economic losses and the number of fatalities from weather- and climate-related events in EU Member States and EEA member countries since 1980.
- The Adaptation Dashboard gives an overview of adaptation aspects at regional level under the EU Adaptation Strategy.
- The European Climate Risk Assesement provides an assessment of current and future climate change impacts and risks relating to the environment, economy and wider society in Europe (published in April 2024).

The Climate-ADAPT case study database gathers and showcases implemented adaptation options and initiatives to create and improve the enabling conditions for adaptation at all governance scales - from local to EU level.

Table 5.2
Topics covered under vulnerability, impacts and adaptation areas of the Global Set Climate Change Statistics and Indicators

Area	Topics
VULNERABILITY	Water security, food security and agriculture
	Vulnerable species, ecosystems and their services
	Buildings and infrastructure vulnerable to climate change
	Vulnerable population
	Area of country vulnerable to climate change
IMPACTS	Agricultural production affected by climate change
	Areas affected by climate change
	Freshwater resources
	Hazardous events and disasters
	Climate change and human health
	Climate change evidence
	Soil condition
	Distribution and status of species
	Distribution and status of ecosystems
	Production and consumption of materials
	Climate change impacts on transport and critical infrastructure
	Climate change impacts on tourism
ADAPTATION	Climate change adaptation policies, strategies and plans
	Risk management, disaster forecasting and early warning systems
	Public awareness of and education on climate change
	Area-based adaptation to climate change
	Climate change monitoring
	Water management
	Waste management

Source: UNSD (2022).

422. **Adaptation policies, strategies, and legislation.** Data and information on assessing the state of play and enabling national environment for adaptation efforts are one relevant area mapped across most of the National Adaptation Plans. To operationalize adaptation actions, national agencies leading on climate change and NSOs will require information to assess and understand the source of their national demands for data on adaptation, the resources allocated to adaptation across different sectors and the legislations in place playing a crucial role to adapt to the effects of climate change. Data and information on policies, strategies, plans and legislation might include number of plans and budget estimations dedicated to climate change adaptation actions across different sectors; list of adaptation actions mainstreamed across the national planning; existence of a coordinating body that contributes fostering adaptation at the country level; information on management measures prescribed in national legislation, policies or multilateral/ bilateral arrangements aiming to reduce direct damage and improve management of ecosystems services in response to changing climate conditions. This might include, for example, fisheries management measures or integrated coastal zone management (ICZM). Identifying the enabling environment for climate change adaptation can give NSOs and their partner line ministries valuable information to initiate a dialogue that can lead to mapping key national stakeholders for adaptation, their roles and data priorities.

423. **Risk and disaster management.** Data related to risk and disaster management are a must when considering climate change adaptation data. Disaster risk reduction national agencies and insurance companies are the most important producers and users of these data.

424. **Climate change adaptation monitoring capacity.** Data from monitoring stations is used in the design, implementation and monitoring of adaptation policies and measures. Many of these data are used for preparing and implementing disaster risk strategies.

425. **Disaggregated data for effective adaptation measures.** Adaptation will be required everywhere but must be prioritized now for the most vulnerable people with the fewest resources to cope with climate hazards. Therefore, a cross-cutting demand for data in the design, implementation and monitoring of adaptation measures includes disaggregated data at the subnational levels including community level, but also by sex, disability groups, ethnicity, etc. to capture and respond to the needs and priorities of communities and the most vulnerable populations. Data disaggregation can also have a more technical lens. For example, as described above, some statistics might require disaggregation also by type of waste treatments and disposals.

426. **Public awareness of and education on climate change.** An emerging area in the data landscape for adaptation is the data and information to track progress on the public awareness of and education on climate change. Data demand for this area on adaptation might include the number of people reached through climate change public awareness, number of companies publishing sustainability reports or number of climate change statistics reports published by national agencies.

427. **Hydrological and hydrometeorological data**, e.g. data on rainfall, streamflow, temperature, humidity, snow water equivalent, snow depth, cloud cover, radiation, evaporation or evapotranspiration, groundwater levels and wind. These data are needed to identify patterns and new trends to anticipate future changes due to climate change and, based on the data, develop and implement the necessary adaptation measures.

428. In the context of adaptation, a key type of data used by policymakers and researchers are **climate projections,** that is, simulated future climate under different scenarios including various emission concentrations and socioeconomic pathways.[82] Although these data sources are not likely to

---

[82] Representative Concentration Pathways (RCPs) are scenarios describing different levels of GHG and other radiative forcings that might occur in the future. Shared Socioeconomic Pathways (SSPs) scenarios describing plausible major global developments based on five narratives describing alternative socio-economic developments, including sustainable development, regional rivalry, inequality, fossil-fuelled development, and middle-of-the-road development. See: Keywan Riahi, et al, The Shared Socioeconomic Pathways and their energy, land use, and greenhouse gas emissions implications: An overview, Global Environmental Change, https://doi.org/10.1016/j.gloenvcha.2016.05.009.

be used directly by NSOs, it is worthwhile to be aware of existence of these data and the types of analyses that users might be employing them for.

5.3 How NSOs can contribute

429. As described in the previous paragraphs, understanding climate change vulnerability and resilience, and planning climate change adaptation are important elements of climate change response (UNECE CES, 2022). Measuring adaptation and vulnerability, however, is still a considerable challenge due to contextual and subjective factors (UNECE CES, 2021).

430. Climate change adaptation is a large and complicated area. Basic scientific knowledge in different areas is needed to understand the data needs, which vary strongly among and within countries, and are context specific. It is also a dynamic area, and it is essential to stay updated about the developments. This complexity is also illustrated with the fact that most of adaptation-related statistics and indicators in the Global Set and the CES Set are at tier 3, meaning lack of internationally agreed methodology and data availability (UNSD, 2023).

431. Although new developments are needed to address that, statistical data from many established domains can also be very useful for informing and monitoring climate change adaptation policies. Moreover, many NSOs already undertake or plan some activities related to climate change adaptation, such as producing relevant statistics, linking and disseminating data from other producers or supporting monitoring of national adaptation plans. They often carry out these activities in collaboration with other national institutions (UNECE CES, 2020).

432. The section below is based on the considerations of NSOs contributing to the work of the Task Force, concerning the role of NSOs in the context of climate change adaptation. The considerations included existing statistical activities which are relevant, existing work which could be relevant with some adjustments, as well as possible future work which could be considered, if human and financial resources are made available. In the examples it is indicated whether the work is already in place or is only planned.

5.3.1 Identifying the current role and coordinating with other organizations

433. As described in the previous chapters, the response to climate change is fundamental for achieving sustainable development. All efforts to achieve sustainability include efforts to address climate change, where climate change adaptation is an important component of the response to the climate crisis.

434. The community of producers and users of climate change-related statistics is increasingly aware of how broad, complex and challenging the measurement of hazards, vulnerability and adaptation can be. Improving measuring adaptation requires collective - global and regional - efforts to develop and agree on concepts, definitions and methods, understand the data needs and share context-specific experiences and responses.

435. Collaboration is key to success in this rapidly evolving area. Building data assets relevant for adaptation policies involves using economic, social and environmental data compiled by NSOs and by a wide variety of agencies and institutions. Through building these assets, NSOs could strengthen their engagement and even take on a coordination role in this area.

436. Many NSOs already undertake or plan some activities related to climate change adaptation, sometimes also in the broader context of sustainable development, or related topic of measuring hazardous events and disasters[83]. A very important opportunity for an NSO to get involved and identify how it can best contribute is to be involved in developing or implementing of the national adaptation plan or strategy (see section 5.2.2).

[83] See also UNECE (2019c).

Australia

The Australian Bureau of Statistics, as Australia's NSO and as a partner in the Australian Climate Service (ACS) - a partnership of Australia's trusted science organizations - has a lead role in providing information to support climate and national disaster decision-making. This includes:

(a) Leading or contributing to components of Australia's first National Climate Risk Assessment (NCRA) which will inform priorities for climate adaptation and resilience actions through the National Adaptation Plan.

(b) Delivering measures and models of the vulnerability of Australian communities to climate change and natural hazards.

(c) Developing a statistical framework to provide a structure for monitoring climate-related indicators - including methods, data sources and gaps.

(d) Enabling improved economic measures of climate risk e.g. through location-based data and integrated land accounts.

(e) Providing leadership and developing capability in the geospatial analysis of information.

(f) Working with data custodians to identify, standardize and integrate administrative data.

Canada

In June 2023, the Government of Canada published Canada's National Adaptation Strategy[84] which identified indicators for adaptation, including seven developed by Statistics Canada (StatCan). As Canada's national statistical office, StatCan is continuing to work with Environment and Climate Change Canada (ECCC), the department responsible for climate change policy, to further develop climate change adaptation indicators and identify data needs under the National Adaptation Strategy monitoring and evaluation framework. The cross-cutting nature of adaptation means that multiple subject matter areas within StatCan have begun work or have an interest in developing work around adaptation, which have been highlighted below.

In December 2020, as part of its strengthened climate plan, A Healthy Environment and a Healthy Economy, the Government of Canada committed to develop Canada's first National Adaptation Strategy with provincial, territorial and municipal governments, Indigenous Peoples, and other key partners. Part of this commitment included the establishment of an indicator framework for measuring adaptation progress at the national level.

In 2021, ECCC started a partnership with StatCan to review existing StatCan data holdings and produce possible adaptation indicators. The objective of this work has been to establish a baseline of readily available climate change adaptation data and identify gaps to guide future development of indicators. In June 2023, ECCC published Canada's National Adaptation Strategy which included 25 targets to support the monitoring and evaluation strategy therein under the key themes of disaster resilience, health and wellbeing, nature and biodiversity, infrastructure, economy and workers, tools and resources, and knowledge and understanding. While some of the proposed targets have data to measure their progress via departmental results, reports from other government departments, some targets are considered novel and require new data to be collected. The Strategy was published

StatCan, as part of the partnership with ECCC, is responsible for:

(a) Researching StatCan's data holdings and developing partnerships with subject matter experts within the organization to understand current and future work;

(b) Researching the climate change adaptation monitoring and evaluation approaches of peer countries and multinational organizations;

(c) Supporting the monitoring and evaluation framework of the Adaptation Strategy by identifying and developing existing and new climate change adaptation indicators for Canada at the national scale; and

(d) Participating in webinars and ad hoc meetings organized by ECCC, as needed.

The cross-cutting nature of adaptation means that multiple subject matter areas within StatCan have begun work or have an interest in developing work around adaptation; where appropriate, these teams will be invited to join

[84] https://www.canada.ca/en/services/environment/weather/climatechange/climate-plan/national-adaptation-strategy.html and https://www.canada.ca/en/services/environment/weather/climatechange/climate-plan/national-adaptation-strategy/full-strategy.html

an internal Working Group which will report to the broader governance structure to be established under the National Adaptation Strategy. The key objectives for the group would include regular program updates, coordination of work, including costing, and identification of data gaps. In addition to providing data for updated and new indicators, StatCan also outlines potential areas for future work. However, funding and resources associated with all the work identified is not guaranteed.

Italy

In the context of the Agenda 2030 for Sustainable Development, since December 2016 ISTAT disseminates information related to the monitoring of the SDG indicators in Italy, updated every six months. Since 2018, ISTAT has been publishing the SDGs report - *Statistical information for 2030 Agenda in Italy* (released also in English since 2019), which aims to guide within the complex system of indicators produced with insights and trend analysis for all Goals, and some interlinkages between Goals.[85]

Referring to Goal 13 "Take urgent actions to combat climate change and its impacts", besides the reporting under 13.2. "Total greenhouse gas emissions per year", inventory totals (UNFCCC) and accounts totals on emissions disaggregated by type of gas, households and industry are disseminated (ISTAT-ISPRA (Italian Environmental Agency)). Under 13.1. "Strengthen resilience and adaptive capacity to climate-related hazards and natural disasters in all countries", statistical measures related to risk, deaths, missing and injured persons for flood and landslides (ISPRA) are disseminated by territorial disaggregation as well as the Global average temperature anomalies over land in Italy (ISPRA), the impact of forest fires (ISTAT, Processing on State Forestry Corps), the Number of the seismic movements by magnitude class (National Institute of Geophysics and Volcanology, National Earthquake) and an indicator "Concern for climate change and greenhouse effect" (ISTAT, Quality of Life Multipurpose Household survey) as a proxy.

The indicators from the SDG database are addressed to public decision makers and used for the National Sustainable Development Strategy, the Ecological Transition Plan, and the national Recovery and Resilience plan, in coordination with national bodies (ministries).

In 2020, ISTAT implemented for the first time the UNECE Set of Core Climate Change-Related Indicators and Statistics Using SEEA at the national level; in 2021 a feasibility assessment of implementation at sub-national level was carried out.[86] Out of 44 indicators, 29 could be implemented for Italy, 13 derived from environmental accounts. For adaptation - recognized as the most challenging area - 3 indicators have been disseminated as proxies or partial indicators related to: Change in water use efficiency over time, Share of green urban areas in the total area of cities, Proportion of agricultural area under productive and sustainable agriculture.[87]

Serbia

Statistics for SDGs. Statistical Office of the Republic of Serbia (SORS) has the role of coordinator of monitoring SDG indicators and cooperates with all relevant statistical data producers and international organizations to provide environment-related SDG indicators. Many institutions of the Republic of Serbia have been actively working and cooperating with the SORS for many years on the production and dissemination of environment-related SDG indicators.

Between 2017, when SORS had only 43 SDG indicators available, for which the data source was SORS, and January 2023, that number on SDG portal has increased to 126 (from 231). Out of 126 available indicators, about 30 are environment-related SDG indicators.

Four national institutions are responsible for 5 of 12 indicators of sustainable development relevant for climate change adaptation, for which data are available on SDG portal. Statistical Office is responsible for SDG 6.4.1 Change in water use efficiency over time and SDG 7.2.1 Renewable share in the total final energy consumption. Ministry of the Interior is responsible for SDG 1.5.3/11. b.2/13.1.2 Number of countries that adopt and implement national disaster risk reduction strategies. The Serbian Environmental Protection Agency is responsible for SDG 12.5.1 National recycling rate, tons of material recycled, and the Ministry for environmental protection for SDG 15.3.1 Proportion of land that is degraded over total land area.

---

[85] See https://www.istat.it/en/well-being-and-sustainability/sustainable-development-goals and https://www.istat.it/en/well-being-andsustainability/sustainable-development-goals/istat-indicators-for-sustainable-development
[86] ISTAT, Economy and Environment, Chapter 13, https://www.istat.it/it/archivio/258752
[87] https://www.istat.it/storage/14-Conferenza-nazionale-statistica/poster/059.jpg
https://www.slideshare.net/slideistat/14a-conferenza-nazionale-di-statistica-250776883

National adaptation process. The Ministry of Environmental Protection of the Republic of Serbia has started the preparation of the Proposal for the Climate Change Adaptation Program during 2022 after the Law on Climate Change was adopted at the end of 2021. They formed a Working group of representatives from all relevant institutions in the country including the Statistical Office.

The document states that it was prepared in accordance with the principles of the European Union Climate Change Adaptation Strategy (adopted in 2021), which mandates the need to implement adaptation as "smart", "systematic" and "quick", with an emphasis on the importance of conservation of water resources, which are considered particularly threatened by climate change.

Domestic experts and members of the academic community contributed to the preparation of the document, bearing in mind that the adaptation must have a scientific basis and is specific to each country or region. The document defines goals, measures and analysis for the following areas: agriculture, forestry, health and safety of people, transport and infrastructure, energy, urban planning and urban development and biodiversity. Water resources are within the framework of the Nexus approach and were not considered separately. As of September 2023, the document is in the process of adoption.

During the work, it was pointed out that the most difficult was to determine the outcome indicators for the defined goals. The sources for outcome indicators are mainly the different reports of the Ministry of the Environment, the Ministry of Finance and web portal Digital Climate Atlas of Serbia created in cooperation with international organizations. Official statistics from various areas are the basis for the analysis in the document. The need for indicators in the field of biodiversity is highlighted as well as data on losses and damages caused by increasingly frequent extreme climatic conditions.

Participation of SORS representatives in the Ministry working group for preparation of the Adaptation Program is extremely important for SORS due to a better understanding of the complex process of adaptation to climate change. In addition, it helped to see the need for new data and indicators in this area that producers of official statistics should include in their future plans.

5.3.2 Identifying and producing relevant statistics and indicators - by statistical domain

437. This section includes both cross-cutting and domain specific examples of statistics and indicators relevant in the context of climate change adaptation. The order of domains does not reflect their importance but approximately follows the order of domains from the Classification of Statistical Activities.

Canada

StatCan has identified several sources of data for monitoring climate change adaptation. These sources have been peer reviewed by other policy departments and have been included in the National Adaptation Strategy. These data sources are:

(a) Percentage of households with air conditioning units (domestic):

  (i) Table 38-10-0019-01 Air conditioners;

(b) Organizations who factored climate change into their decision-making process:

  (i) Count of public organizations who factored climate change adaptation into their decision-making process, by core public infrastructure assets, Infrastructure Canada;
  (ii) Count of municipal organization who factored climate change adaptation into decision-making process, by core infrastructure assets, by urban and rural, and population size, Infrastructure Canada;

(c) Organizations with a documented asset management plan:

  (i) Asset management practices for publicly owned core infrastructure assets;
  (ii) Asset management practices for municipally owned core infrastructure assets, urban and rural, and population size;

(d) Capital expenditures on environmental infrastructure:

  (i) Capital expenditures, non-residential tangible assets, by type of asset and geography (x 1,000,000);

(e) Direct payments to agriculture producers:

(i) Direct payments to agriculture producers (statcan.gc.ca);

(f) Crop data:

(i) Estimated areas, yield, production, average farm price and total farm value of principal field crops, in metric and imperial units (statcan.gc.ca);

(g) Green spaces in urban areas:

(i) Urban greenness and normalized difference vegetation index by 2021 population centre.

Other sources of data have been identified by StatCan, but have not been peer reviewed yet, and thus are not included in this section.

France

In France, the statistical department in charge of producing, disseminating and developing statistical information in the areas of climate change, environment, energy, transport and construction/housing is the Data and Statistical Studies Department (SDES) of the General Commission for Sustainable Development (CGDD) within the Ministry for Ecological Transition and Territorial Cohesion. SDES is part of the French official statistical system as well as the European Statistical System (as an "Other National Authority" (ONA)).

In 2017, the SDES produced an indicator of exposure of individual houses to the shrinkage-swelling of clays by municipality[88], at the request of the General Directorate for Risk Prevention, and as part of its participation in the National Observatory of Natural Risks. The indicator was updated in 2021, to take into account new hazard zoning. The indicator reflects the exposure of individual houses to the phenomenon of clay shrinkage-swelling with regard to their location on the territory of municipalities in mainland France.

In 2020, SDES analysed the exposure of population, property and environment to climate related risks[89]. The analysis showed that in 2016, climate risks potentially concerned six out of ten French people. The vulnerability of territories depends on the exposure of different sectors of the local economy, and strong disparities between territories can be observed in this regard. The vulnerability assessment addressed several types of phenomena: floods, landslides, forest fires, atmospheric phenomena and avalanches, and showed that that most of the French territory is subject to some of these risks. The exposition of the population was categorized as very strong, strong, medium, safe, very safe and without climate risks.

Ireland

Many projects and activities related to climate change adaptation in the Central Statistics Office (CSO) can be used to establish adaptation strategies by identifying levels of exposure and vulnerabilities among the Irish population. However, more could be done to publish data monitoring and evaluating already ongoing adaptation.

Flood management. Flood management is becoming more important in Ireland in terms of climate change adaptation, with numerous parliamentary questions on this topic being submitted to the CSO in recent years. The Irish Office of Public Works (OPW) has 'responsibility for leading and coordinating the implementation of the National Flood Policy which involves the development of a planned programme of feasible works, with a greater emphasis on non-structural flood risk management measures.' The CSO participates in the Office of Public Works' annual National Flood Management Conference but has yet to look into using Office of Public Works data to monitor these implementations.

Other topics. Exploring topics relevant for climate change adaptation, on which some new statistics could be produced, the CSO Ireland has compiled the following list of potentially relevant topics:

(a) Coastal protection

(i) Erosion threatening buildings near coastline
(ii) Sand dunes protection
(iii) Strengthening harbour walls

---

[88] https://www.statistiques.developpement-durable.gouv.fr/nouveau-zonage-dexposition-au-retrait-gonflement-des-argiles-plus-de-104-millions-de-maisons?rubrique=43&dossier=200

[89] https://www.statistiques.developpement-durable.gouv.fr/risques-climatiques-six-francais-sur-dix-sont-dores-et-deja-concernes?rubrique=43&dossier=200

(b) Disasters and extreme events

- (i) Data needed to inform assessment of damage e.g. population along coastline or in a mountain valley
- (ii) Location of access points such as bridges
- (iii) Food
- (iv) Changes in our diet arising from changes in food availability due to lower yields etc.

(c) Hydrology gauges

- (i) To measure river flow levels - high levels indicate flooding risk and low levels indicate risk to fish life etc.

(d) Invasive species

- (i) Arising from a warmer, wetter, climate
- (ii) Pest and weed control
- (iii) Change in seed varieties or livestock breeds to those more resistant or suitable
- (iv) Reintroduction of predator species such as eagles and wolves
- (v) Protection of native species
- (vi) Seed banks

(e) Meteorological data

- (i) Extend time series to improve understanding of exceptional weather

(f) Phenology

- (i) Changes in timing of bird and insect migration changes
- (ii) Relocation of fish and wildlife to locations with more suitable climate in winter and summer

(g) Repair and prevention strategies

- (i) Reforestation after forest fires or windblow damage
- (ii) Replanting of coral reefs if damaged by warmer seas
- (iii) Forest fire breaks

(h) Tourism

- (i) Introducing new activities such as mountain biking if snow season shortens
- (ii) Artificial snow

(i) Water supply

- (i) Fixing water leaks in anticipation of water supply shortages
- (ii) Repairing water reservoirs
- (iii) Water metering and water charges

5.3.2.1 Demographic and social statistics

438. As defined by the IPCC (see definitions of impacts in section 5.2.1), relevant adverse consequences include those on lives, livelihoods, health and well-being, economic, social and cultural assets. Consequently, the statistical domains related to demographic and social statistics, such as population, migration, labour, education, health, income and consumption, social protection, human settlements and housing, culture, time-use, and others can contribute providing primary data, developing analyses and assessments of vulnerability, exposure and risk for certain population groups and social dynamics. More methodological and research work needs to be developed with respect to these areas.

Box 5.3
Integrating gender into climate change-related statistics

Integrating gender into climate statistics

Environment statistics is one of the statistical fields traditionally considered gender-neutral and often developed without much consideration of the different impact that environmental conditions have on the lives of women and men owing to existing gender inequities (United Nations, 2016). Increasingly it has been acknowledged that understanding the nexus between gender and the environment is crucial to understand social and environmental inequalities and barriers to sustainable development as well as to provide opportunities for transformative actions (UNEP and IUCN, 2018). The lack of data disaggregated by sex and gender statistics is one of many factors that render women and girls and their needs and priorities invisible to policymakers. Filling data gaps in the gender-environment nexus, including with regard to climate change and disaster risk reduction, is thus a key step in informing effective, evidence-based policies and programmes to achieve gender equality (UN Women, Women Count, and United Nations (2022).

UNSD Global Set and gender

At the 51st session of the United Nations Statistical Commission, in 2020, the global statistical community "Requested that a gender perspective be adopted and integrated into all the agenda items of the Commission, as is, for example, the current practice in the Statistical Conference of the Americas" (Decision 51/115, b). In response to this decision initial efforts at integrating a gender perspective into the work of the Commission were undertaken by the Statistics Division. To support the process and review, the Inter-Agency and Expert Group on Gender Statistics offered to take on the task of investigating how to integrate a gender perspective across the work of the Commission, and highlighted the challenges it carries, particularly when making the case for mainstreaming a gender perspective into all statistical areas, including in economic and environmental pillars and climate change for developing specific examples and guidance. Work on the climate and gender nexus started in 2022 by the Expert Group on Environment and Climate Change Statistics (EG-ECCS) and is currently being advanced by a sub-group established under the EG-ECCS to look at climate change questions from censuses and surveys. The sub-group is engaged in discussions with the IAEG-GS on mainstreaming gender into climate at the national level.

5.3.2.2 Health statistics

439. As explained earlier in the chapter, there are significant climate-change risks related to human health. Through their work on health statistics, statistical offices can produce data that can be used for both monitoring observed climate change impacts in this area and assessing how successful the introduced adaptation measures are compared to a baseline scenario.

Italy

For example, data on causes of death provide information on mortality patterns and form a major element of public health information. Causes of death data refer to the underlying cause which - according to the World Health Organization (WHO) - is "the disease or injury which initiated the train of morbid events leading directly to death, or the circumstances of the accident or violence which produced the fatal injury". The information provided in the medical certificate of cause of death is coded into the International Classification of Diseases codes according to the rules specified in the Classification. The climate crisis has increased the average global temperature and is leading to more frequent high-temperature extremes, such as heat waves. Higher temperatures can cause increased mortality. The most vulnerable members of the population, such as the elderly and infants, will be most severely affected.

Although they may seem to be two very different issues, climate change and air pollution are closely interlinked. For instance, air pollution in the form of particulate matter from diesel engines is circulated around the globe, ending up in the most remote places, including the polar regions. An interesting study conducted in Italy concerns how exposure to pollution can affect mortality in both urban and rural areas. For this purpose, a few years ago, ISTAT started a collaboration with the Department of Epidemiology, Health Service of the Lazio Region/ASL Roma 1, to analyse some evidence on the differential effects of short-term PM exposure on all-cause mortality throughout Italy and by urban, suburban and rural administrative contexts.

A new domain explores how temperature changes over a given time period might alter or influence mortality by cause. For this purpose, daily all-cause and cause-specific mortality data could be analysed; for example, it might be interesting to assess the potential increased risk of death from accidents due to extreme weather events (floods, fires and storms). In addition, mortality analysis can be related to the main potential sources of social inequalities, such as income or education, to assess whether socioeconomic level can modify the effect of climate change on mortality.

The source related to the Demographic Balance data of the resident population provides information on the flows of demographic dynamics acquired through the notifications sent by the municipalities; the monthly detail allows to observe how the concentration of excess mortality in some recent years characterized by heat waves, is higher in the summer months and in the elderly ages.[90] In 2022, one-third of the excess mortality over the expected value is concentrated in the months of July and August, which were characterized by an abnormal heat wave. [91]

Canada

Health Analysis. In the fall of 2022, the Health Analysis Division published the analytical paper *Exploring the associations between cooling centre accessibility and marginalization in Montreal, Toronto, and Vancouver, Canada*[92]. In this paper, "the potential spatial accessibility of cooling centres within a 15-minute walk was measured to analyse the associations between cooling centre access and marginalization."

The division also conducted a study linking socio-economic impacts to the prevalence of air conditioning in Canada. Using data from both the Household and Environment Survey (HES) and the Canadian Community Health Survey, the main objectives of the study were to quantify and describe patterns of household air conditioning rates according to socio-demographic characteristics and health conditions and to estimate household air conditioning rates amongst heat-vulnerable populations. This research was published in the summer of 2023: *The prevalence of household air conditioning in Canada*[93].

Number of deaths due to climate change. StatCan maintains and disseminates data from the Canadian Vital Statistics – Death Database. This database contains "...basic indicators (such as counts and rates) on deaths of residents of Canada. Information from this database is also used in the calculation of statistics, such as cause-specific death rates and life expectancy". It may be possible to link this database to the Canadian Coroner and Medical Examiner Database to further refine estimates for cause of death linked to climate change.

Air quality – indoor and outdoor. StatCan currently has several datasets related to indoor and outdoor air quality that stem from the HES. This survey includes questions on awareness of air quality advisories, changes in behaviour and impacts on health to members of the household. The HES is a biennial survey, but these series of questions are not available for every reference year.

United Kingdom

A four-year project has been established, led by the UK Office for National Statistics (ONS)[94] in collaboration with the African Institute of Mathematical Sciences, Rwanda; the University of Ghana; UK Health Security Agency; and Cochrane climate-health working group (University of Alberta, Canada), financially supported by Wellcome[95]. The project will advance global research on climate and health, help to address gaps in the knowledge base and support national monitoring and evidence-based climate adaptation policies. It will produce a detailed, harmonized framework of standards and methods, primarily for national statistics offices (NSOs), incorporating multidisciplinary and multilateral inputs as well as building capacity for defining and reporting climate and health statistics. The intention is to produce a series of climate and health interaction metrics that will have clear unified definitions, in a way that is both similar and complementary to the global frameworks such as the UN Global Set, specifically on development of a transparent and generalizable framework for official statistics on climate change and health.

[90] https://www.istat.it/it/files/2023/03/Dinamica-demografica2022.pdf
[91] ISTAT. Annual Report 2023 https://www.istat.it/it/archivio/286191
[92] Health Analysis Division published the analytical paper Exploring the associations between cooling centre accessibility and marginalization in Montreal, Toronto, and Vancouver, Canada https://onlinelibrary.wiley.com/doi/10.1111/cag.12805
[93] *The prevalence of household air conditioning in Canada* https://www150.statcan.gc.ca/n1/pub/82-003-x/2023007/article/00002-eng.htm
[94] https://www.ons.gov.uk/aboutus/whatwedo/programmesandprojects/standardsforofficialstatisticsonclimatehealthinteractions
[95] Raphael Sonabend. Standardising health and climate metrics to drive urgent action. 2 February 2022. https://wellcome.org/news/standardising-health-and-climate-metrics-drive-urgent-action

The project is also developing a global reporting and knowledge-sharing platform and open-source toolset to facilitate high quality research and official statistics reported by NSOs, in order to help achieve national climate objectives and report regular official statistics on climate and health interaction. Finally, the aim is to explore statistical methods to better estimate climate-related health risk using real world data sources, including novel and big data, and modelling local impacts in order to inform decision makers for better climate adaptation policies. The open-source tools and framework as well as global guidance will be shared through an online knowledge sharing platform which will be hosted by the UN Global Platform team (in collaboration with ONS) for wider accessibility to multidisciplinary users, stakeholders, funders and policy makers.

The exposure to climate-related hazards can include biological, chemical, or physical stressors and can differ in time, locations, populations and severity. These are referred to as exposure pathways (Figure 5.6). Climate change threats may also accumulate over time, leading to longer-term changes in resilience and health. The climate related health risk topic areas were identified based on a scoping review, the IPCC AR6 report and areas that WHO has identified as health risks (2023).

Figure 5.6
Climate-related hazards and health - exposure pathways

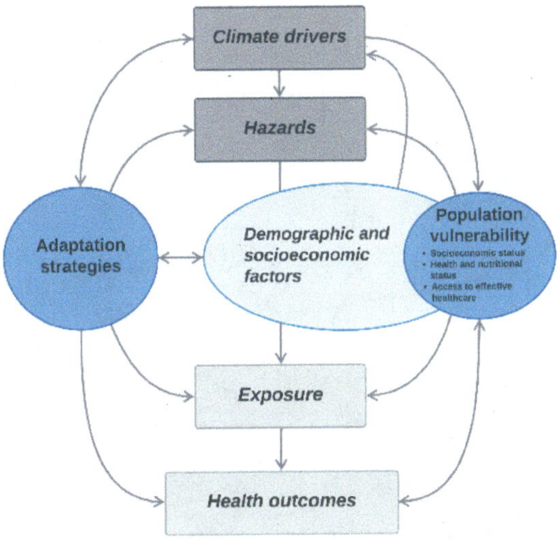

Source: ONS.

Table 5.3
Climate hazards and their health impacts identified based on a scoping review, the IPCC AR6 report and areas that WHO has identified as health risks

Climate hazards	Health impacts
Extreme weather events (e.g. flooding, wildfires, drought, storms)	• Injury and mortality from extreme weather events
Heat and cold waves	• Heat- and cold- related mortality • Morbidity (e.g. hospital admission)
Air pollution	• Respiratory illnesses (e.g. respiratory tract infections) • Non-communicable diseases (e.g. cancer, cardiovascular diseases)
Increasing temperature and precipitation, extreme weather events	• Water-borne diseases and other water-related health impacts (e.g. cholera, and other gastro-intestinal) • Zoonoses (e.g. avian flu) • Vector-borne diseases
Climate variability (e.g. drought and permafrost thaw)	• Malnutrition and food-borne diseases
All climate hazards	• Mental and psychosocial health • Effects on health systems

Source: IPCC AR6 report and WHO

Figure 5.7
Highest and lowest temperatures in England, by taking the hottest and coldest 5% of days over the entire period from 1988 to 2022

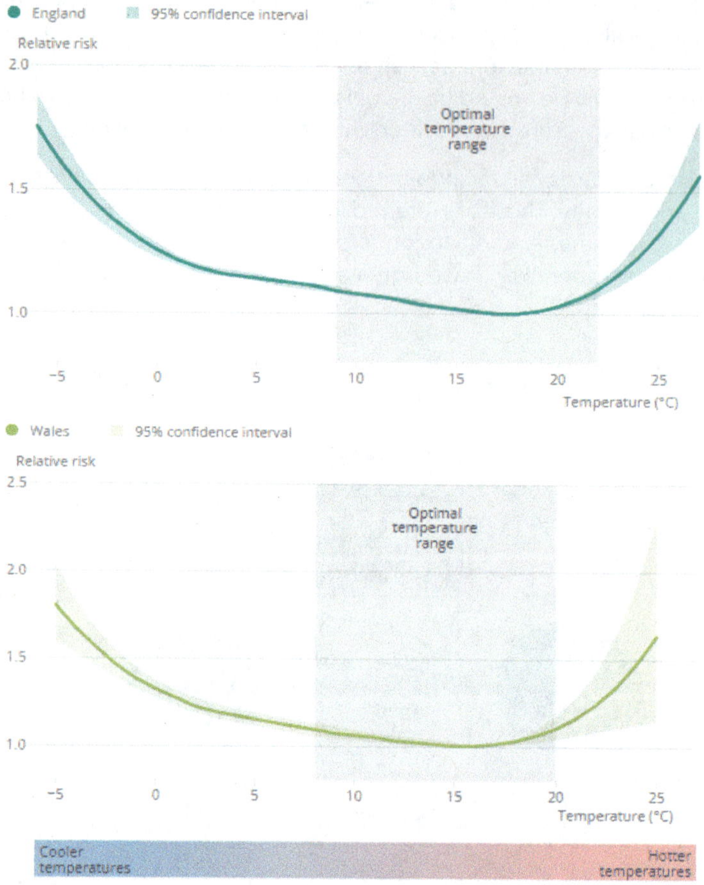

Source: ONS.

Climate-related mortality estimates, England and Wales - Publication of the Office for National Statistics[96]. The daily time series of mortality from 1988 to 2022 was combined with climate information from the Centre for Environmental Data Analysis (UK Meteorological Office[97]). A statistical method was developed to understand the impact of temperature on mortality risk, while adjusting for other factors influencing mortality. Optimal temperature ranges are presented as part of this analysis, to highlight temperatures at which mortality risk was lowest. The methods developed in this project produced the analysis which estimated and considered the highest and lowest temperatures in England, by taking the hottest and coldest 5% of days over the entire period from 1988 to 2022 (Source: IPCC AR6 report and WHO

Figure 5.7). These extremes corresponded to temperatures below 2 degrees Celsius for the coldest days, and above 18 degrees Celsius for the hottest days, in England overall. We calculated the deaths attributable to these temperatures to understand the potential deaths that could be avoided by limiting exposure to very high or very low temperatures. These methods were developed and applied in the Wellcome-funded project contributing to standardized heat and cold related metrics.

5.3.2.3 Economic statistics

440. Anthropogenic impacts arising from economic systems can be limited by decreasing pressures; at the same time, economic systems are at risk from the consequences of climate change. In addition to expenditures for adaptation, many other areas of economic statistics are relevant to

[96] Gillian Flower, Victoria Masding, Emily Blissett. Climate-related mortality, England and Wales: 1988 to 2022. Office for National Statistics, 22 September 2023. https://www.ons.gov.uk/peoplepopulationandcommunity/birthsdeathsandmarriages/deaths/articles/climaterelatedmortalityandhospitaladmissionsenglandandwales/1988to2022
[97] https://www.ceda.ac.uk/

adaptation, among them the issues of transportation and tourism, which can be affected by the consequences caused by climate change, as well as more sustainable solutions that can reduce the causes and vulnerability.

Canada

Climate Change Mitigation and Adaptation Expenditures. StatCan is currently assessing the feasibility of responding to 14 different recommendations to fill international gaps in data identified by the International Monetary Fund (IMF) and approved by the G20 countries. Seven of the recommendations identified as part of this Data Gaps 3 Initiative (DGI-3) are related to climate change, with the estimation of current and capital expenditures for climate change mitigation and adaptation being of particular interest to a national adaptation strategy.

Clean technology. StatCan tracks the use of clean technologies related to environmental protection, sustainable resource management or adapted goods by enterprises, and makes these data available on their website as well as on the Open Government platform. However, the distinction between technology for mitigation and adaptation is not clear.

Tourism. StatCan measures tourism from multiple angles, including number of trips, employment and economic impact. Tourism, as most sectors of the economy, will be impacted by climate change. Seasonality within the tourism sector and the incongruent weather-related impacts of climate change on this seasonality make it challenging to measure overall adaptation within the sector. Measuring the overall impacts on tourism will, therefore, require detailed data at levels of disaggregation that do not currently exist within the StatCan databases.

5.3.2.4 Agriculture statistics

441. Agriculture statistics is a very important area of work for many statistical offices. Agricultural production is closely related to food production, and food security is at risk, as the vulnerability of this sector to droughts, floods and other consequences of climate change is increasingly evident as well as need for climate change adaptation.

Canada

Agricultural water use. StatCan, through the Census of Agriculture and the Agricultural Water Survey, publishes several datasets related to water-use and irrigation within the agriculture sector. However, these data are published only for farm operations that irrigate and are not normalized with respect to total cultivated area or production. Therefore, some work is required before direct applications to adaptation can be made.

Food security. StatCan has several datasets related to food insecurity some of which pertain to Indigenous peoples. These datasets measure food insecurity as affordability, or the ability to pay for food, rather than availability of food. Climate change may have an impact on the price, and therefore affordability, of foodstuffs.

Harvesting activity of Indigenous groups. The Indigenous Peoples Survey runs every 5-years and is a national survey on the social and economic conditions of First Nations people living off reserve, Métis, and Inuit. The questions on harvesting activity speak to frequency of the activity and subsequent activities resulting from harvesting. The survey could potentially be relevant but at this point it would be difficult to ascertain a link to climate change using existing questions.

Serbia

SORS collects and publishes data on primary agricultural production and change of its volume, as well as data on fishing. For the needs of calculation of physical volume index of agricultural production, used are the data from regular surveys in scope of agricultural production on produced quantities of crop and livestock production, and data on average prices of agricultural products' producers on the level of the Republic of Serbia. The list includes all economically significant agricultural products.

The most often used SORS agriculture statistics[98] in analyses related to adaptation and the negative impact of climate change are data on arable land, crop production (yields according to agricultural crops), fruit growing,

[98] https://data.stat.gov.rs/?caller=SDDB&languageCode=en-US

especially viticulture fruit production, data on meadows and pastures, livestock production, and data on irrigation by agricultural crops and organic fertilizers.

Italy

ISTAT is producing the most part of statistical indicators related to agriculture and forestry in Italy. Basically, almost all data produced satisfy EU Regulations and are finalized to provide information on agriculture and forestry production and how farmers and forestry enterprises organize their work and use production tools to guarantee their revenues. There are no projects devoted to the evaluation of climate change effects on agriculture and vice-versa. However, data collection concerns issues that may be related to climate change.

The most important indicators related to climate change are: 1) number of livestock per municipality; 2) irrigation; 3) agricultural surfaces and livestock managed through organic farming; 4) use of pesticides; 5) wastes of crop production; 6) surfaces occupied by forests and forest removals. An in-depth analysis of how and how much these indicators can affect climate change is missing.

The most important gap is the lack of data sources - for instance, a specific survey - concerning the strategies adopted by agricultural holdings for facing the effects of climate changes on their production, and for reducing the negative effects of agricultural practices on climate change.

5.3.2.5 Energy statistics

442. Energy is a wide and complex domain with many facets that make this area extremely relevant not only to mitigation but also to adaptation. It is an area in which the two types of response need full integration, to develop synergies and avoid trade-offs. Statistics related to energy have a decisive role in relation to the definition of mitigation and adaptation interventions.

Ireland

CSO makes use of administrative microdata to produce statistics relevant to climate change adaptation. Examples include Building Energy Ratings based on imputed energy consumption. The Sustainable Energy Authority of Ireland maintains a database of all energy rating audits. The CSO publishes these on a quarterly basis. The data show trends in main space heating fuels, dwelling type, and total floor area. New dwellings and dwellings being sold or rented are required to have a Building Energy Rating audit undertaken. The most notable trends, since the series began in 2009, are much more energy-efficient new dwellings and a change to using electricity as the main space heating fuel. Energy efficiency is linked to SDG 7.3 (Double the global rate of improvement in energy efficiency by 2030) and SDG 7.2.1 (Renewable share in the total final energy consumption).

Italy

The ISTAT Household Energy Consumption Survey provides a picture of the energy endowments possessed by Italian resident households, how they are used, and their efficiency based on the seniority of the systems, in a sector that is constantly evolving technologically, toward more efficient systems. The survey also aims to investigate the most virtuous behaviours of citizens and their degree of awareness with respect to their contribution to energy conservation. These issues are relevant to the vulnerability of certain segments of the population, being at risk of energy poverty, and ability to convert to more efficient systems, in the residential sector. The ISTAT Survey on Enterprises Final Consumption of Energy Products (soon to be implemented) is aimed at providing data on final energy consumption of the various energy sources in the industrial and tertiary sectors for monitoring the achievement of the objectives of the Integrated National Plan on Energy and Climate to reduce consumption and improve energy efficiency in the industrial and tertiary sectors.

5.3.2.6 Environment statistics

443. Climate change and more in general anthropogenic activities burden natural systems and territories with pressures that alter the balance with the environment and undermine the regenerative capacity of natural resources as well as the conservation of ecosystems. Global warming heavily affects the hydrological processes toward an intensification of the water cycle with cascading effects that can vary in space and time: droughts and water scarcity on the one hand, and intense and concentrated precipitation on the other.

444. Availability and quality of water in all sectors is at risk: civil, agricultural, industrial and energy uses, as well as the risks of disaster events. Statistical information on hydrometeorological data, on water use and all environmental issues enables the analysis of pressures, expected impacts, risk assessment and adaptation, in all different contexts.

Ireland

Water Statistics. The CSO carries out the FAO Aquastat survey which is linked to SDG 6.4.1 (Change in water use efficiency over time) and SDG 6.4.2 (Level of water stress). The CSO input the water parameters in the questionnaire and the SDGs are automatically calculated. However, there is insufficient data with regards to water use efficiency as, with exceptions such as some potato crops, irrigation is less common in Ireland relative to other countries.

With regards to Wastewater Treatments, the CSO is currently at a data collection stage for the OECD/Eurostat Joint Questionnaire on Inland Waters. Work is still in early stages, and it is not yet confirmed if it will lead to a release or publication.

Forestry Statistics. The CSO publishes data on Afforestation Area which analyses the area afforested under the Department of Agriculture, Food and the Marine afforestation schemes (for the period 2007 to 2021 at time of writing). [99]

Other releases of Forestry statistics include Forest Wood Removals, based on the annual CSO Roundwood Removals Survey which was designed to collect data on roundwood and biomass removals from privately-owned forests in Ireland.

Climate Data Rescue[100]. The CSO has a role in the development of the Irish Statistical System including ensuring that the statistical potential of administrative records is realized, and so the CSO offered to cooperate with Met Éireann, Ireland's meteorological office, in the rescue of pre-1960s daily meteorological data.

Inconsistencies with formatting and handwritten records prevented straightforward scanning of documents without extensive quality checks. Given the CSO's experience with the labour-intensive clerical work involved in capturing and processing large scale datasets, the CSO would take on the task of digitizing and quality checking the data in need of rescuing. The data in question comes in the form of monthly weather sheets, some dating as far back as the 1820's, containing an average of 40 variables of daily data each, plus summary monthly tables. The variables include such thermometer and barometer readings, rainfall, humidity, wind force, etc. There are character variables, such as wind direction, and also general remarks on phenology, unusual weather events, timing of storms, etc.

A longitudinal file for each meteorological station will be made available to Met Éireann by the CSO for integration with data from 1960 onwards already computerized, before eventual publishing as a research file online. As the file will be used on a daily level, the quality of the daily data must be reliably accurate. Thus, the CSO carried out extensive checks for consistency, verifiable totals and outliers.

Statistics on extreme weather events and disasters helps policymakers to understand the types, frequencies and intensity of hazards in a given climate, and thus the risk of adverse consequences facing human and ecological systems as a result of climate change. Developing successful adaptive measures requires statistics on such hazards. The advantages of daily data such as those rescued allow for granular analyses to identify the frequency of unusual and extreme weather events such as heatwaves or absolute droughts over an extensive time period. Daily data allows users to carry out flexible calculations of indicators. For example, the rescued data allows for an analysis of the change in the growing season in Ireland, i.e. the frost-free period in a year. An increase in the length of the growing season could have beneficial effects on crop growth but also detrimental effects on biodiversity with the presence of pests in previously colder months.

This climate data rescue project is only possible with the co-ordination between the NSO and other agencies, in this case the CSO and Met Éireann, in finding the statistical potential of administrative microdata. While Met Éireann will make the research files of fully integrated daily data available publicly, the CSO will publish our own climate indicators report using the integrated time series. The report should show a thorough picture of Ireland's historical and present climate, and the change in patterns of our extreme weather events.

[99] For more details on the 2014-2020 scheme, see Afforestation Grant and Premium Scheme 2014-2020: https://assets.gov.ie/69781/6f6dc37d067d4e7a813605acadd6b77e.pdf
[100] https://www.cso.ie/en/methods/climateandenergy/csoclimatedatarescue/

Italy

Meteoclimatic and hydrological data. Since 2017, the ISTAT Survey of Meteoclimatic and Hydrological Data is annually carried out by the Meteoclimatic, Other Events and Natural Resources Statistics Unit, collecting daily measurements on air temperature and precipitation from about 500 certified meteorological stations located on the ground (unit of analysis), mostly placed within urban areas of the main Italian cities. Stations are managed by several Italian Institutions, Research Centres and Authorities (both public and private), respondents to the Survey. Due to their vulnerability and exposure to climate change effects, urban systems are defined climatic hotspots by international scientific Institutions. Moreover, cities are the centre of adaptation policies to face climate change, needing highly detailed territorial data for analysis and monitoring targets. A georeferenced ISTAT Meteo Database has been developed, collecting geographical and technical data on the stations, to produce yearly meteoclimatic statistics and Indices of extremes by provincial municipality.

The ISTAT Meteo Database has been regularly updated (time series 1971-2021), properly storing daily data on minimum, mean and maximum temperature, total precipitation level and humidity degree by station. 21 Indices of extremes are produced by 109 provincial municipalities, expressed in physical units (C°, mm, number of days), in absolute and relative values. The Climatic Normal 1971-2000 and 1981-2010 have been calculated for each city, to compare indicators' annual values with the corresponding average climatic values of reference periods, which allows for calculating climatic anomalies indicators and understanding climate variability and climate changes in the cities. Some indicators are part of contextual indicators provided by ISTAT for the SDG 11 for Italy.

The availability of complete series of daily data by station is a condition for developing consistent statistics and indices, calculating climatic anomalies, describing phenomena and their trends in the medium and long run. According to available time series by city, annual Indices are compared with CLINO 1971-2000 and 1981-2010 average values for Regional Capitals and with 2006-2015 average values for Provincial Capitals. Many efforts have been undertaken to strengthen data collection, extend the historical series of daily data available for cities, search for new sources, and to improve timeliness of data produced. Recent publications are available on the ISTAT website.[101]

Water statistics and accounts. ISTAT is engaged in the production and updating of water statistics and water accounts. As the country suffers the lack of a national information system on water resources, ISTAT has been actively working for several years to meet users' needs, fill data gaps and provide high quality statistical information, in line with the national and international requests. Following are some key activities performed in these years.

Water balance. An ongoing joint project between ISTAT and Italian National Institute for Environmental Protection and Research (ISPRA) aimed to provide a homogeneous assessment on the state of water resources and the indicators derived (e.g. Water Exploration Index plus, SDG 6.4.2) to international (Eurostat, EEA, OECD, FAO, etc.), national, regional and local bodies.

Public water supply. Through the Urban water census, ISTAT collects information on water withdrawal for drinkable use (35,000 abstraction points), adduction and supply networks, public sewage (at municipal level) and urban wastewater treatment (more than 18,000 plants). Data collection started in the 1950s, but it has been consolidated in the current survey since the 1999 edition. Respondent units are the water operators (approximately 2,400 in 2020), both water utility companies and local authorities. The survey contents have been progressively updated by considering emerging information needs. Since 2018, the survey frequency has been reduced to two years. The last census referred to the year 2020. Data are elaborated at several territorial levels: e.g. administrative aggregations, river basin districts, water bodies, sewage agglomerations.

Water use in agriculture and industry. To overcome the lack of an adequate monitoring system for non-civilian uses, ISTAT has developed, tested and validated a methodology for estimating water volumes in both agriculture and industry. As a result of these estimations, made at the national, river basin district, regional and local labour

[101] Recent ISTAT publications include:

- Temperature and precipitation in Italian regional and provincial capital cities Years 1971-2021 ISTAT Statistical Report May 2023 https://www.istat.it/it/archivio/284549
- Blue Book 2023 Utilitaties-ISTAT-ISPRA Report for the World Water Day March 2023 https://www.utilitatis.org/my-product/blue-book-2023
- Climate Change Statistical Measures, Year 2020, ISTAT Statistical Report March 2022 https://www.istat.it/it/archivio/268615
- Water Statistics, ISTAT Statistical Report for the World Water Day March 2023 https://www.istat.it/it/archivio/282387

system levels, the total water withdrawal and use for public water supply, industry and agriculture has been calculated.

Wastewater statistics. The number of urban wastewater treatment plants in operation is high and there is a partial validation of wastewater quality parameters because of missing data. Some critical issues in the calculation of the total volume of wastewater generated and treated, at the national and subnational level, are due to the limited knowledge on plants for exclusive industrial use. There is much to be done to ensure a continuous, regularly updated, timely and comprehensive production of water statistics. The improvements in existing surveys and the evaluation of data integration with administrative archives are further steps needed.

To build citizens' awareness on sustainable use of water, ISTAT provides an annual focus for the World Water Day, including text, data tables, infographics and maps.[102]

Hydroelectric sector. Effects of climate change on hydrological cycle also affect the production and availability of energy in the hydroelectric sector, which is very important in Italy among renewable energy sources. As part of adaptation actions, this sector needs a revamping of infrastructures because of their obsolescence.

Serbia

According to the annual plan and program of statistical surveys, SORS conducts the following five annual surveys in the field of water statistics:

(a) Survey on water use and protection against pollution which provides data on the quantities of abstracted, used and wastewaters from industry section;

(b) Survey on drinking water supply, which provides data on water abstraction, overtaken water from other networks and water provided to other systems; water distributed to users, water used for own consumption, and total water losses; data on water treatment and water supply network;

(c) Survey on urban wastewater which provide the data on quantities of wastewater discharged in the recipient; quantities of discharged treated and untreated wastewater; sewerage system/ network;

(d) Survey on irrigation which collects data for water abstracted and used as well as irrigated areas by type of irrigation;

(e) Survey on protection against damaging water effects which collects the data regarding protection against damaging water effects classified into four topics: Flood protection and rivers management; Land drainage; Land erosion protection and flood control and Consumption of fuels.

The data from the mentioned survey are mostly available to users in the SORS database[103] in area environment-water as well as in the regular annual publication called Eco-Bulletin. SORS participate in international data collection and regularly submits data on water statistics to Eurostat through the Joint Questionnaire on inland waters, as well as to FAO through the AQUASTAT questionnaire.

Example of uses of SORS data in topic on climate change adaptation. In the Second National Communication[104] of the Republic of Serbia under the United Nations Framework Convention on Climate Change, table 6.2 Adaptation measures to changed climate conditions in the hydrology and water resources for the strategic area Water use, one of the measures for Increasing the efficiency of the water supply system includes reduction of losses to an optimal level.

SORS is producing an indicator on water losses in distribution annually and by municipality from the survey on drinking water supply.

Droughts because of climate change in the Republic of Serbia are becoming more frequent, which can result in water shortages in certain areas of the country and reduced flow of smaller rivers, and thus impaired water quality. The key sources of pollution are municipal and industrial wastewater, which seriously impairs water quality and river ecosystems. Population connected to wastewater treatment with at least secondary treatment is an important indicator calculated from SORS annual survey on urban wastewater.

Natural hazards and climatic events (floods, snow drifts, and high temperatures) also affect the road infrastructure. Floods that trigger landslides and damage to drainage systems have the most damaging impact. Data from the

[102] UNSD, Envstats News and Notes, December 2022, https://unstats.un.org/unsd/envstats/Newsletter/Issue52.pdf
[103] https://data.stat.gov.rs/?caller=SDDB&languageCode=en-US
[104] https://www.klimatskepromene.rs/wp-content/uploads/2017/09/SNC_eng.pdf

SORS survey on Protection against damaging Water Effects also provides data on flooded roads and roads protected against floods in kilometres, which SORS publishes in the Statistical Yearbook and Eco-Bulletin.

5.3.2.7 Hazardous events and disasters

Disaster-related statistics

445. Climate change adaptation relates closely with hazardous events and disasters, such as floods, droughts, fires and heatwaves etc., the frequency and magnitudes of which are expected to increase. Therefore, in many countries such data may have been collected by disaster risk management agencies or similar institutions. There is a thematic overlap between hazard/ disasters risk management and adaptation to climate change since many disasters are defined as climate related and climate change is expected to exacerbate them.

446. Besides disaster risk management agencies, other adaptation-relevant data are often collected by specialized government agencies (other than NSO) as suggested above. NSOs need to thoroughly examine and assess such work to avoid duplication in case new data collection is undertaken. They also need to determine where they can help, usually in terms of improving data quality, validating and disseminating. The starting point should be the country's NAP and disaster risk reduction strategies.[105]

447. As presented in section 2.3.8, a recently developed CES Set of Core Disaster Risk-Related Indicators includes the core indicators on measuring disaster risk recommended to be produced. See also Box 5.1 earlier in this chapter discussing some terminological differences between the climate change and disaster risk reduction domains.

Canada - Disaster Readiness and Emergency Preparedness

The 2023 iteration of Household and Environment Survey will add content related to natural disasters and emergency preparedness, including questions on whether and how a household was impacted by a natural disaster and how prepared a household is for an emergency.

In addition, the Canadian Centre for Justice and Community Safety Statistics has a project team responsible for the development of data and analysis of the risks and impacts of natural disasters to various populations and communities. This work has included a pilot community fire risk reduction dashboard[106] and there are plans to develop a socio-economic dashboard for emergency preparedness and response.

StatCan also has an Emergency and Disaster Tiger Team, that acts as a quick connect to the network of analysts who can provide data for client requests about a climate change related event, most recently supporting the Atlantic provinces during Hurricane Fiona. The team requires funding to continue this service that will support, coordinate and ensure cohesion in any data provided in response to the request, and conduct a post-mortem to support effective provision of these data in a more general manner.

5.3.2.8 System of Environmental-Economic Accounting

448. Ecosystem Accounting (EA), through data on extent and conditions of ecosystems, as well as on ecosystem services (ES) they provide and the connected economic values (rents, actual and potential costs, etc.), may provide useful information for adaption policies, through the integration of ecological and socioeconomic information. ES assessment is useful for identifying the most favorable ecosystem types and conditions, for quantifying needs, for evaluating investment needs, costs and physical benefits, and for monitoring and evaluating policies.

449. The methodology in this area is still evolving. Data needs include, in addition to traditional population and socioeconomic data, data from registries, Big Data, Earth observation and citizen science. Substantial investments are also needed in the methodological and conceptual aspects of the matter.

[105] OCHA, List of Disaster Management Authorities, https://reliefweb.int/national-disaster-management-authorities
[106] https://www.statcan.gc.ca/en/sc/video/community-fire-risk-reduction

Canada - Census of Environment

The Census of Environment is a new program by StatCan that will track the size and condition of ecosystems, including forests, agroecosystems, wetlands and urban areas. While still in early stages of development, the Census of Environment is developing ecosystem accounts and relevant environmental statistics following the System of Environmental-Economic Accounting-Ecosystem Accounting (SEEA-EA) statistical standard. Selected datasets released to-date include ocean and coastal ecosystem extents, protected and conserved ocean and coastal ecosystem extent, saltmarsh condition and services, agroecosystem extent and condition, renewable water yield and urban greenness. In addition, work is ongoing to develop terrestrial land cover and ecosystem extent accounts. These accounts will help deliver indicators for reporting on Global Biodiversity Framework targets, among other initiatives.

Italy

In Italy, EA is being developed in collaboration with the environmental agency (ISPRA) and feeds the annual report of the Committee for Natural Capital. Climate change reduces certain ES and makes other more necessary than ever. The estimates of some ES are based on the study of the spatial intersection between elements at risk (lives, activities, assets) and the potential supply of ES, as well as the probability of the occurrence of adverse events.

For example, the ES "flood protection" requires a good degree of soil permeability and runoff resistance in locations uphill of settlements and infrastructure, so that exceptional rain events do not cause their flooding. The estimation of this ES may consider both potential and *actual supply* of this service. Similarly, "Air cooling" is an ES that will be estimated, consisting in the *Protection* from *Heat Islands*, provided by urban green areas of great relevance in terms of adaptation, whose estimation could usefully be coupled with data on the quantity of people benefitting from it and their socioeconomic vulnerability.

5.3.3 Linking data from different sources or domains

450. In addition to being a provider of key datasets, NSOs can also play a significant role in helping to unleash the analytical value of these datasets through linking data and publishing analyses combining data from various domains. The example below presents an analysis of data from several sources put together to answer complex policy questions about the efficiency of climate response and its impact on the economy - also showcasing the connection between climate change mitigation and adaptation.

Ireland - Decoupling Emissions from Economic Activity[107]

In 2023, the CSO developed a new statistical release to show the progress in reducing GHG emissions towards Ireland's 2030 Climate Action Plan targets. The release focuses on the five sectors with the highest GHG emissions and examines the economic importance of those sectors in terms of gross value added and employment, as well as the economic and social impacts of reducing emissions towards the targets.

The Decoupling Emissions from Economic Activity release combines data from environmental-economic accounts such as Air Emissions Accounts with National Accounts data such as gross value added and employment numbers. The report incorporates analysis of use figures from Supply and Use Tables, data on the agricultural labour force and livestock units, vehicle data, and social and environmental data such as household energy poverty rates, household electricity consumption and heating fuels, and environmental subsidies to households.

An important aspect of the release is monitoring of the impacts of reducing emissions on particular economic sectors. The release addresses the question of whether emissions and economic activity are decoupling, or if the reduction in emissions results in decreased employment and economic activity in these high-emitting sectors. Furthermore, data from Supply and Use Tables are used to evaluate interactions between economic sectors. For example, the significant use by the food manufacturing sector of agricultural outputs shows the potential knock-on impact on this sector as emissions from agriculture are reduced.

[107] CSO, Decoupling Emissions from Economic Activity, https://www.cso.ie/en/statistics/environmentaccounts/decouplingemissionsfromeconomicactivity/

5.3.4 Subnational focus and geospatially enabled statistical data

451. As explained in section 5.2 of this chapter, the right adaptation solutions depend on the local context. Therefore, statistical data describing the local context with sufficient granularity, which can be collocated with climate observations and projections are necessary to inform the decision making and measure the success of implemented approaches.

452. Especially relevant in this context are population and housing censuses, which offer the possibility of capturing data with the highest possible granularity.

453. NSO can contribute to this aspect through:

(a) Collecting and disseminating data about specific types of localities, e.g. cities, municipalities and urban areas; coastal areas; mountainous regions;

(b) Examine the possibility of including questions related to climate change in population and housing censuses or producing releases showcasing the value of the census data for climate analysis and policy;

(c) Consistent georeferencing of its statistical outputs;

(d) Collaborating with geospatial authorities on improving interoperability of the statistical data with geospatial data.

Italy[108]

The Urban Environment Survey is carried out annually by ISTAT and collects environmental information related to main Italian cities, with the aim of providing useful information to monitor the quality of the urban environment and support policies to ensure health and wellbeing for all in cities. The survey consists of eight modules related to Air, Eco management (that includes water rationing), Water (measured in coordination with ISTAT Water Census for civil use), Energy, Mobility, Waste, Noise and Urban Greening. In particular the information is related to: diffusion of air pollutants and measures to contain; production of energy from renewable sources, efficient use of energy, certification and energy upgrading of buildings; initiatives to prevent the production, facilitate proper delivery and collection of urban waste; noise measures, exceedances of limits and temporary activities with significant noise impact; demand and supply of local public transport, sustainable mobility and info-mobility; publicly managed urban green, actions for the development of green spaces and protected natural areas; and planning tools related to major urban environmental issues. Aimed at 109 provincial capitals and metropolitan cities through the network of municipal statistical offices with the Integration of NSS sources of administrative data on mobility, energy, urban waste, environmental certifications eco management, the survey collects 500 primary data, which are used to produce 180 indicators.

The survey has been conducted since 2000 in support of the DPSIR (Driver-Pressure-State-Impact-Response Framework) and it produces relevant information for climate change adaptation, vulnerability and exposure in the context of cities. Indicators from this survey are used within the ISTAT Reporting on SDG, Goal 11, for example for targets 11.6.2 indicators on concentrations and exceedances of several air pollutants, 11.2.1 Seat-Km of public transport networks and for 11.7.1 the 'Incidence of urban green areas on urbanized area of the cities', as a proxy indicator. Urban indicators are also relevant for the National Sustainable Development Plan, the Ecological Transition Plan and the National Recovery and Resilience plan. Data from this survey have also been integrated with temperature and precipitation data from the ISTAT Survey on Meteoclimatic and hydrological data, to analyse the relationship between climate variables, drivers of climate change and pollution, and consequences in terms of extreme weather events.

An exercise in georeferencing data related to noise could also be extended to other relevant issues, complementing the ISTAT work related to the project of census data at sub-municipal level. An emerging issue concerns electric vehicle recharging, for example. Strengthening collaborations with other agencies, such as environment agency and ministries is essential, also for strengthening administrative data with high potential.

[108] ISTAT, I Cambiamenti Climatici: Misure Statistiche, 2022, https://www.istat.it/it/archivio/268615
ISTAT, Documents on Urban Environment, https://www.istat.it/it/archivio/ambiente+urbano
ISTAT, Principali Fattori Di Pressione Sull'Ambiente Nelle Città Italiane, 2018, https://www.istat.it/it/files//2021/01/ebook-principali-fattori-pressione-ambiente.pdf

Recently, to enhance the statistical information available on green areas and to increase the quality of the disseminated statistical data, a new line of activity was introduced: acquisition via open-source software 'Lymesurvey' of the geographical data of green areas, in format (.shp - shapefile), containing the following information of the green area: name, address, size in square meters and management (public/private).

Census Maps. Coordination of geographic data production activities related to Census Maps, project created for the collection of population census data and its dissemination at the sub-municipal level. With the transition to the permanent census in Italy, Spatial Bases have remained fundamental as a geographic layer for the dissemination of sub-municipal data in compliance with privacy regulations. The Spatial Bases are a geographic structure for disseminating any georeferenced data at the sub-municipal level from any survey. Each polygon is classified according to a particular code that specifies the anthropogenic and natural characteristics of the area; characteristics that are extremely useful for the assessment of risks, due to climate change and land use. In Italy climate change is the cause of adverse weather phenomena and natural disasters with disastrous multi-risk cascade events: landslides, floods, forest fires, extreme weather phenomena, heat waves, drought, desertification, water shortages, etc. Italy is also subject to volcanic phenomena and earthquakes. Fragility, vulnerability, poor land management, maintenance and obsolescence of infrastructure aggravate human, economic and environmental losses and damages.

Different indicators relevant for climate change adaptation can be extracted from the geographic data of Enumeration Areas such as:

(a) Increase in inhabited areas over the years;

(b) Vulnerability of public buildings and infrastructure points (ports, airports, main roads and railways) in the territory;

(c) Assessment of the population that is vulnerable and most exposed to the risk of environmental disasters, including those caused by climate change.[109]

GIS Web MapServices and Applications. The activities include:

(a) Publishing Web Maps including geospatial and statistical geo-referenced layers using Italian and European standards

(b) Publishing the geo-metadata on the Italian National Repertoire for Territorial Data to make those data discoverable and interoperable on the web

(c) Using the maps and layers integrated with other data sources available on the web or inside the NSI, to perform spatial analyses, such as: climate changes analysis, such as comparison and dispersion models, affected areas, etc.

(d) Development and the maintenance of the website https://gisportal.istat.it/[110]

(e) Using basic GIS tools such as spatial queries and overlay.

National Urban Audit involves data harmonization process which is carried out with the Commission and other Member states to organize and disseminate comparable data on European cities. The harmonization work is based on sharing methodologies and variables definitions as well as on the production, selection and dissemination of the data with common tools and objectives. the following domains relevant to climate change are covered by the dataset:

(a) Demography (data on the population by sex and by age group for cities and functional urban areas)

(b) Transport (data on the total amount of vehicles in cities - there are plans to publish data on vehicles by type of motor energy in the future)

(c) Environment (Length of bicycle network)

(d) Waste (Total amount of urban waste)

---

[109] Mugnoli S., Abbatini D., Chiocchini R. Lipizzi F., ISTAT Cartography as a support for spatial emergencies, Geomedia, 2016, https://mediageo.it/ojs/index.php/GEOmedia/article/view/1363

[110] Examples of interactive WebGIS applications are available on Istat website showing georeferenced statistics on the maps. Data are aggregated at admin unit level (NUTS, LAU) and other functional and statistical units (enumerations areas, grids). https://gisportal.istat.it/IstatViewer/?locale=en

Climate Effect Atlas in the Netherlands[111]

The Climate Effect Atlas was created in 2007, when several provinces and research institutes felt the need to provide low-threshold access to national climate information. The atlas has been managed by the Climate Adaptation Services Foundation since 2012. In 2020, the Ministry of Infrastructure and Water Management commissioned the foundation to update and expand the Climate Effect Atlas and improve the viewer. The current version is available since May 2021. Climate Impact Atlas is primarily intended for professionals who are directly or indirectly engaged in climate adaptation. However, the Atlas can also be used by other interested parties.

Figure 5.8.
Example from the Climate Effect Atlas in the Netherlands - drought risk stress.

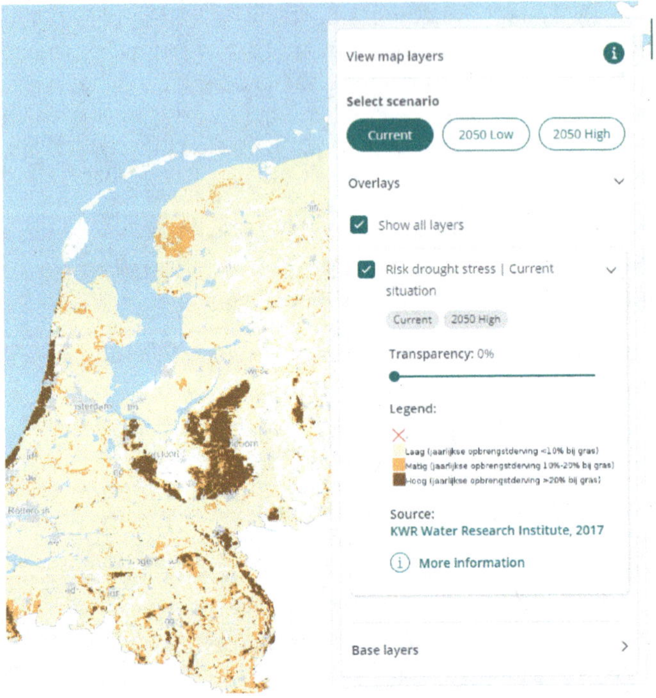

***Source*: Climate Effect Atlas in the Netherlands**

Upon visiting the Climate Impact Atlas, the user first enters the Viewer, where they will find all the maps with a brief explanation. The maps in the Atlas can be used to gain a picture of the physical consequences of climate change in an area, and how this will impact residents and the environment. The maps can also help to place the issue of climate adaptation on the political agenda. The map narratives provide background information to the main map layers in the Viewer and assist the user in utilizing the maps. Furthermore, the map narratives relating to waterlogging, drought and heat also feature diagrams showing various climate statistics.

Although Statistics Netherlands is not actively involved in the Climate Effect Atlas, this tool is an example of a good practice of addressing data needs at a local level.

5.3.5 Sharing data and microdata for research and risk analyses

454. As explained in section 2.2.1, many NSOs provide access to microdata for researchers in strictly defined and regulated conditions. This function is particularly important in the context of climate change adaptation, where many knowledge gaps persist and research on applicable policy measures is lacking (see

[111] https://www.klimaateffectatlas.nl/en/

455. Table 5.1). Although NSOs can also produce analytical outputs linking various data sources and statistical products at high levels of granularity, the scope and variety of information needs related to climate change adaptation is vast and context specific, and many questions may refer to future and projected climate developments, which is not what statistical offices usually get involved in. It is, therefore, very important to make sure that researchers and analysts who can examine these questions have the possibility to tap into the wealth of data assets available at NSOs.

456. With the technological advancement allowing the statistical offices to make microdata easier to access and use (in equally safe and secure manner) and the researchers to undertake increasingly complex microsimulations and modelling, the relevance of microdata access can be expected to increase in the future.

For example

The FAO Economic and Policy Analysis of Climate Change programme of the FAO Agrifood Economics Division:

- Uses household survey data to understand factors that influence the productivity and resilience of food production systems in the context of climate change including program impact, farm-level management and technology adoption decisions, and the role of institutional, geographic and household characteristics in influencing these;
- Combines farm level socio-economic data, macro-economic scenarios and climate models to detect the heterogeneous impact of climate change on the economy, employing an integrated approach building on the interplay between macro-economic models such as Computable General Equilibrium models and micro-simulation models;
- Merges available geospatial data on soil quality and historical climate tends with household survey data to better understand the effects of climate change and variability on farmers 'adaptive responses and to identify factors that may enhance welfare (e.g. yield, income, food security, nutrition etc.) or welfare stability under conditions of climate shocks.

5.3.6 Playing a role in adaptation monitoring and evaluation

457. As explained in section 5.2.2., a continuous monitoring and evaluating (M&E) is an integral part of the NAP process. A data plan for the NAP M&E should be developed and supported with adequate resources and data should be collected and stored in a systematic manner. It has also been recommended that the responsibility for compilation and aggregation of data collected from multiple agencies should be clearly assigned to one agency, which could be the national statistical office. Undoubtedly, NSO would be well placed to play a role in data compilation and aggregation, as well as participate in developing indicators used in the M&E framework.

458. The UAE Framework for global climate resilience adopted in Dubai at COP28 already includes quantitative and time bound targets around the adaptation cycle. NSOs can play an active role in reporting for the targets. Indicators for the thematic targets around water, food, health, ecosystems, infrastructure, poverty and cultural heritage will be further elaborated in a 2-year UAE - Belém work programme (2023 to 2025), relying to the extent possible on existing data and indicators under the SDGs or Sendai Framework, as well as identifying gaps and developing new indicators where required. Involvement of NSOs in this process would ensure the feasibility of producing indicators nationally, allow for leveraging existing data collections and help NSOs prioritize the work on statistics related to adaptation.

5.3.7 Promoting harmonization and international comparability through use of existing classifications and indicators

459. Developing an exhaustive, internationally agreed set of climate change-adaptation indicators is challenging, due to, among others, number of dimensions to consider, diversity of circumstances, priorities and success definitions. However, a number of statistical guidelines and classifications is

already in place that can ensure that even though the selection or prioritization of indicators may depend on national priorities, the indicators produced follow the same methodology and are comparable among countries and in time, and not produced in an ad-hoc manner. This can significantly increase the value of statistics and indicators for policymaking and research as well as ensure credibility and trust.

460. Commitment to international standards and international collaboration in statistics are part of NSOs DNA, as expressed in the Fundamental Principles of Official Statistics (see Box 2.2) and can contribute the knowledge of classifications, frameworks and standards to the national process of identifying and producing data for climate change adaptation.

5.4 Conclusions and recommendations

461. As discussed in section 5.2, climate change adaptation can be understood as the process of minimizing adverse impacts of climate change by reducing exposure and vulnerability and increasing adaptive capacity and resilience. From an NSOs perspective, providing statistics for adaptation policymaking does not exclusively involve producing data and statistics on the adaptation measures alone - the information about the exposure and vulnerability and adaptive capacity and resilience is just as important. Adaptation response requires information on not just the hazards themselves but also the level of exposure in a given country to them, the vulnerabilities of the population and economy, and the resilience present to cope with hazardous events. Together, these elements inform the level of risk in a society for adverse consequences of climate change and of the human response to climate change. The adaptive measures are changes in processes, practices and structures intended to reduce the levels of climate change related risk through the elements of exposure and vulnerability, and thus lessen the final adverse impacts.

462. To adequately inform adaptation responses, NSOs should understand the scope of information needed for effective climate change adaptation policy making. A useful tool for achieving this is the NAP process, which enables identifying country's adaptation needs and for creating strategies and programmes to address these needs.

463. The *2022 Synthesis report for the technical assessment component of the first global stocktake, State of adaptation efforts, experiences and priorities* mentions the key challenges related to adaptation faced by countries. While many are related to limitations in data availability and information, there is mention of challenges related to insufficient coordination among the main stakeholders that hold climate data and thus have a role to play in helping societies adapt to climate change. To identify key stakeholders, NSOs could prioritize based on criteria such as the ability of the stakeholder to identify, produce and share data that is relevant to the country's interests and decision-making and reporting processes.

464. Central to data and statistics related to climate change adaptation is the conceptual and statistical understanding of climate-related hazards in terms of their occurrences, the exposure and vulnerability of and eventual impacts to population and ecosystems. In this regard, the disaster risk reduction (DRR) community has as part of its core expertise long dealt with these measurement and data collection issues pertaining to extreme events and disasters where climate-related hazards are a critical part. For instance, the Inter-Agency and Expert Group on Disaster-Related Statistics (IAEG-DRS) under the auspices of the United Nations Statistical Commission is in the process of developing a global common framework on disaster-related statistics aimed at establishing an internationally coherent statistical reference and guidance document for disaster-related statistics. Additionally, the United Nations Office for Disaster Risk Reduction (UNDRR) and the International Science Council (ISC) have conducted a review of hazard definitions and classification to support the monitoring and tracking progress of the Sendai Framework for Disaster Risk Reduction 2015-2030, the Paris Agreement and the Sustainable Development Goals.

465. Hence, the climate change community in its effort to support informed decisions on climate change adaptation would benefit from engaging the DRR community to build on existing knowledge

and expertise. It will as well provide the opportunity to promote greater alignment in the production and use of climate change and disaster-related statistics. Subject to specific national contexts, NSOs could play a key role in facilitating a closer collaboration between the climate change and DRR communities both at the national and global levels.

466. Country examples help to illustrate that since there is no 'one-size-fits-all-solution' that can apply to the unique context of each community, business, organization, country or region, it is the responsibility of the NSO to contribute to the work of achieving effective adaptation policies that best address these unique contexts. This can be done through applying and improving activities already undertaken by NSOs to the work related to climate change adaptation such as coordination with other organizations, producing relevant statistics across many statistical domains, linking data from different sources, focusing on the need for enhanced geospatial data, etc. As with most statistical products, NSOs should address the need for adaptation solutions relevant to a society's context by prioritizing the demands of users for statistics that are of optimal use for their purposes.

467. Going forward, NSOs should heed the guidance included in the 2023 COP28 decision on reporting for the thematic targets under the UAE Framework for Global Climate Resilience as there is an opportunity to play an active role in the reporting process. The indicators for measuring progress achieved towards these "area" targets (on water, food, health, ecosystems, infrastructure and human settlements, poverty eradication and cultural heritage) and "process" targets (on the adaptation cycle) are due to be elaborated upon in the 2023 – 2025 UAE – Belém programme, carried out jointly by the Subsidiary Body for Scientific and Technological Advice and the Subsidiary Body for Implementation.

468. Based on these conclusions, the following recommendations are made:

(a) Data needed to understand climate change adaptation and reduce climate change related risks are not exclusive to data on adaptation measures themselves. Elements that feed into these risks, namely climate related hazards, exposures to them, vulnerabilities and resilience must also be measured. NSOs can play a critical role in an information system that holistically addresses adaptation by bringing together available information and identifying and filling the gaps;

(b) To help in understanding the scope of information required for effective adaptation policy making, NSOs should inform themselves on the progress of their country's NAP process and, where possible, they should become involved in the process;

(c) NSOs should collaborate more closely with the disaster risk management community and producers of disaster-related statistics;

(d) NSOs should also consider the following avenues of improvement of relevant data:

(i) NSOs should coordinate with other key stakeholders and establish their roles as part of the task in achieving effective climate change adaptation policy;

(ii) Each NSO should prioritize user demands to determine which adaptation statistics are of most relevance to the context of each country. It is recommended to use tools and frameworks that are readily available, for example, by making use of the CISAT or by mapping statistical products to Figure 5.1 as an illustration;

(iii) NSOs should begin identifying and producing new relevant statistics where possible;

(iv) NSOs should review existing collections of data and adjust accordingly to suit user demands for adaptation statistics where possible;

(v) Improvements should be sought in georeferencing data for territorial analysis, perhaps by adding different layers of information in different statistical domains;

(vi) NSOs should seek to improve the availability of microdata for sharing with researchers since knowledge gaps are still present and an increase in demand to address these gaps can be expected going forward;

(vii) NSOs should take advantage of available guidance such as that included in the 2023 COP28 decision on reporting for the thematic targets under the UAE Framework for Global Climate Resilience.

Chapter 6 Statistics for just transition policymaking

6.1 Introduction

469. In addition to a transition in the physical environment, the climate crisis also requires a social transformation. While the impacts of climate change affect everyone in society, the most vulnerable populations can face compounded risks as a result of lack of adequate climate action.

470. The biggest climate change impacts, such as more frequent extreme weather events leading to food insecurity and water scarcity, will be felt most by vulnerable groups. Additionally, communities already struggling with poverty may have fewer resources to adapt to climate change and/or to take mitigation measures. According to IPCC, mainstreaming effective and equitable climate action will not only reduce losses and damages for nature and people, but it will also provide wider benefits (IPCC, 2023b). To ensure that no one is left behind, it is essential that this transition happens in a fair and equitable way.

471. A just transition refers to the equity dimension of the transition towards net-zero emission economies and climate-resilient societies, to achieve global climate goals. To ensure an equal distribution of costs and benefits, as well as a participatory and inclusive transition, the socioeconomic dimensions of environmental impacts and actions need to be considered. Data can support a just transition by enabling policy makers to identify vulnerable population groups and socioeconomic challenges associated with climate impacts and responses. This can include identifying people from marginalized groups who cannot shift to a low-carbon economy, changes in the structure of the labour market due to the disappearance of jobs in fossil fuel industries, or the measurement of energy poverty.

472. NSOs are traditionally responsible for production and dissemination of socioeconomic statistics and (micro)data, including data on population characteristics and their living conditions, household income, and energy statistics, which provide information on fossil fuel use, household consumption or energy prices, and gender statistics. Important data sources are the Survey on Income and Living Conditions (e.g., EU-SILC[112]) and various household surveys. Cross-analysing this data with climate change data and statistics is key to monitoring the social implications of climate impacts and policies, and target solutions towards vulnerable and marginalized groups.

473. This chapter aims to help NSOs to unpack the policy terms linked to just transitions and relate them to their daily work, so that social equity and inclusion in the context of climate change can be mainstreamed into official statistics to advance towards a just transition. As part of future work, the idea of a 'glossary' of policy-related terms could be explored to facilitate communication between policymakers and the more neutral statistical community. Such an exploration begins with attempts to provide a definition for the umbrella term 'just transition'.

6.2 Policy context and definitions

Why do we talk about a just transition?

474. As described by IPCC (2023b), climate change impacts are felt differently by different population groups between and within countries, with those who have historically contributed the least to human-caused climate change being often the most vulnerable and disproportionately affected by its consequences. Today, the biggest impacts of climate change are being significantly felt by parts of Africa, Asia, Central and South America, and Small Island Developing States (SIDS); as well as by small-scale farmers, women, children, indigenous communities, and low-income households across the globe. Hence, there prevails an unequal distribution of the adverse impacts of climate change and related losses and damages to natural ecosystems and people across systems, regions, population groups, and economic sectors.

[112] https://ec.europa.eu/eurostat/web/microdata/european-union-statistics-on-income-and-living-conditions

What is a just transition?

475. IPCC (2022b) defines 'just transitions' as:

> *"A set of principles, processes and practices that aim to ensure that no people, workers, places, sectors, countries or regions are left behind in the transition from a high-carbon to a low-carbon economy.*
>
> *It stresses the need for targeted and proactive measures from governments, agencies, and authorities to ensure that any negative social, environmental or economic impacts of economy-wide transitions are minimised, whilst benefits are maximised for those disproportionally affected.*
>
> *Key principles of just transitions include respect and dignity for vulnerable groups; fairness in energy access and use, social dialogue and democratic consultation with relevant stakeholders; the creation of decent jobs; social protection; and rights at work.*
>
> *Just transitions could include fairness in energy, land use and climate planning and decision-making processes; economic diversification based on low-carbon investments; realistic training/retraining programmes that lead to decent work; gender-specific policies that promote equitable outcomes; the fostering of international cooperation and coordinated multilateral actions; and the eradication of poverty.*
>
> *Lastly, just transitions may embody the redressing of past harms and perceived injustices (ILO 2015; UNFCCC 2016b)"*

476. "Leave no one behind" is a central, transformative promise in both 2030 Agenda for Sustainable Development and the EU Green Deal[113]. It covers social equity and inclusion aspects. The concept of a "just transition" has also been broadly defined as "an economy-wide process that produces the plans, policies and investments that lead to a future where jobs are green and decent, emissions are at net zero, poverty is eradicated, and communities are thriving and resilient" (Just Transition Center and B Team, 2018).

477. Hence, a just transition is a broad term encompassing a system-wide transformation across social, economic and environmental systems, promoting change in two dimensions: 1) emissions reductions and 2) fair, inclusive sustainable development, across national, regional, and global scales. On the one hand, a just transition is a conversion towards low carbon economies that meets the net zero target by 2050, set by the Paris Agreement. On the other hand, it is a transition to societies that thrive through sustainable development pathways, while adapting to climate change impacts and integrating measures to reduce emissions (IPCC, 2023a). In addition, the underlying premise for a just transition is that an environmentally sustainable economy, which is net zero and nature-positive, will benefit all and provide a more equitable future (World Wildlife Fund for Nature, 2021).

478. To enable an equitable society-wide transformation embedded in the just transition idea, there are three crucial enabling factors to consider: 1) recognition, 2) participation and 3) fair distribution of benefits and burdens (IPCC, 2023a and Bennet et al., 2019).

479. Recognition involves acknowledging diverse actors involved, allowing to identify differentiated contributions, vulnerabilities, values and knowledges related to climate change (see also section sections 2.1.2 on stakeholders and 7.2.2 on users of climate change-related statistics).

480. Meaningful participation entails inclusive governance, that is, the inclusion of vulnerable and marginalized communities most affected by climate impacts in the political debate and decision-making processes to guarantee a proper representation of their needs and tasks (see section 7.2.1 for more information about participation).

[113] https://commission.europa.eu/strategy-and-policy/priorities-2019-2024/european-green-deal/finance-and-green-deal/just-transition-mechanism_en

481. Finally, a fair distribution of benefits and burdens generated by climate change adaptation measures and mitigation actions to reduce emissions aims to balance winners and losers through compensation measures, to avoid burdening the most vulnerable.

Why is data key to enable a just transition?

482. A just transition is deeply embedded in the global 2030 Agenda for Sustainable Development, in particular SDG 13, which focuses on sustainable climate change mitigation and adaptation efforts that address key vulnerabilities and promote resilience through a nature-based and people-centred approach. Therefore, for achieving a successful just transition, countries aim to place people's well-being at the heart of climate solutions to leave no one behind. As populations are diverse and have varying needs, timely, accurate, and granular climate change-related data are crucial to understand how the complex environmental, social and economic linkages to climate change-related transitions affect societies, and especially the most vulnerable, in different ways.

483. Nowadays, proliferating climate change-related data and statistics from public and private sector actors not only unlock new solutions, but also accentuate coordination, governance and data quality challenges. Adopting a holistic approach is key to ensuring that data enable governments to understand and respond effectively to heterogeneous populations' needs. High quality climate change-related (micro)data, when used and shared effectively, facilitate monitoring and analysing impacts of climate change, assessing risks and vulnerabilities, anticipating the consequences, guiding investment decisions, and ultimately, encouraging informed and inclusive actions from private and public sector actors that respond to the needs of those most impacted by climate change.

6.3 The role of NSOs and how can they contribute

484. NSOs can act as the key data provider to support global, regional, and national policy activities to ensure a worldwide just transition as regions and countries gradually move towards net-zero emissions economies and climate-resilient societies. In this independent and impartial key role, NSOs provide timely, accurate and granular (micro)data not only on socioeconomic and sociodemographic aspects, but also on living conditions (e.g., EU-SILC).

485. By providing disaggregated data, NSOs can contribute significantly to the development and implementation of just transition policies that are truly inclusive and equitable. Disaggregated data allows policymakers to design interventions that specifically support the most vulnerable groups, ensuring that no single community bears a disproportionate share of transition costs. This is especially important in regions and sectors most directly impacted by the shift away from fossil fuels. Cultural considerations can also receive more emphasis, as in jurisdictions with indigenous populations a generic approach may be insufficient.

486. With detailed socioeconomic data, NSOs can better measure the effectiveness of just transition policies, which can be used to adjust strategies where necessary. This ensures that the policies are not only fair in theory, but also in practice, with real-time adjustments based on ongoing data analysis. Since a just transition is fundamentally about equity, detailed data are essential for understanding the diverse impacts on different demographic groups. This understanding is critical to ensure that the benefits of climate policies are distributed in a fair way and that support mechanisms are in place for those most affected.

487. Such NSO data can be used, for example, to map transition risks of mitigation and adaptation policies, at urban and rural level, and to identify the (socially) vulnerable groups who suffer the most from these risks, such as low-income households, pensioners, orphans, and disabled people. To respect their privacy, the NSOs are officially obliged to regard the collected individual data as strictly confidential, which means that the data on individuals (or companies) may only be used for statistical purposes and that rules and measures must be taken to prevent disclosure of their data.

488. To harmonize regular data compilation, one of the main roles of NSOs is to establish standards regarding definitions, methodology, and indicator development. NSOs can therefore contribute to just transitions by their dissemination of official statistics on demographics, environment, living conditions, consumption, energy use, income, labour, gender, and other topics that fall under regular statistical programmes of NSOs. This regular data flow can be expanded through a further exploration of administrative data and/or by conducting extra surveys to gather additional data, e.g., on impacts and vulnerabilities.

489. In collaboration with other stakeholders NSOs can develop harmonized indicators to measure progress towards a just transition. Together with other stakeholders NSOs have an important role to ensure that all data is used effectively, regarding the costs (no duplication), in solving the data gaps, and in linking the microdata from different data sets, while respecting the confidentiality. One of the key challenges: a lot of data already exists today, but there is a need for better cross-analysing data and data sharing. To increase public awareness for inclusive climate action, NSOs must also develop a different language to reach all social groups. This is not easy, and a lot of poor-quality information is already reaching vulnerable groups.

490. Four case studies below illustrate the role of NSOs in advancing just transition: (1) impact of carbon tax on fossil fuels (Luxembourg), (2) Domestic Building Energy Ratings from social perspective (Ireland), (3) the Dutch Energy Poverty Monitor (the Netherlands), and (4) identifying vulnerable groups (Armenia). These four case studies mainly focus on the impact on household well-being. In principle, transition measures also have impacts on economic sectors, e.g., a ban on fossil fuels implies the closure of coal mines (energy sector), which could also lead to cascading impacts in non-energy sectors. To avoid higher unemployment, affected economic sectors must find ways to provide their workers with new skills and alternative employment.

Luxembourg: Impact of a carbon tax on fossil fuels

In order to meet its commitments under the Paris Agreement and mitigate the GHG emissions responsible for climate change nationally and abroad, the Government of Luxembourg has decided to introduce a carbon tax on fossil fuels sold in Luxembourg. Aware that introducing a new tax on certain energy products could constrain low-income households and increase fuel poverty, the Government commissioned a study[114] at STATEC (National Institute of Statistics and Economic Studies) to establish the amount of tax needed to have a significant effect on reducing GHG emissions, and to identify a method of gradually redistributing the amount of the tax to households. This study proposed the introduction of a tax level of 20 EUR/tCO2 for the year 2021, followed by a gradual increase of +5 EUR/tCO2 for the years 2022 and 2023. It also put forward a proposal for a tax credit that could be introduced to offset the effects of the tax and have a neutral impact on the budgets of low-income households (see Figure 6.1).

Figure 6.1
Tax credit offsets CO2 tax for lower incomes in Luxembourg

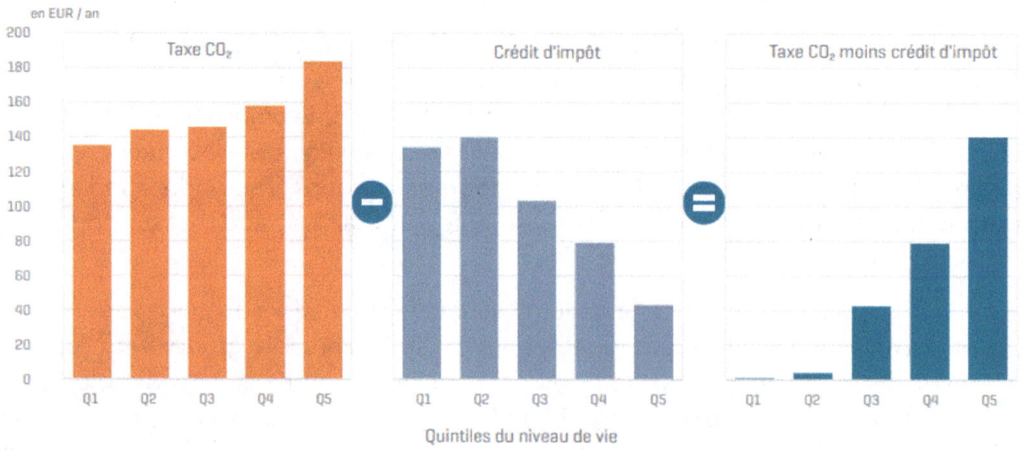

Source: STATEC

114 https://statistiques.public.lu/fr/publications/series/analyses/2020/analyses-08-20.html

Ireland: Domestic Building Energy Ratings from a social perspective

The Central Statistics Office (CSO) published the report Domestic Building Energy Ratings from a Social Perspective[115]. This combined Census of Population and energy audits to look at who is living in poorer quality housing and hence perhaps in need of more financial assistance to upgrade their home to reduce heating emissions.

CSO Ireland has recently combined data from their Survey on Income and Living Conditions with energy ratings and electricity consumption. The main idea was to exclude high quality housing from energy poverty even if the household spent over 10% of disposable income on energy costs (which could be e.g. heating a swimming pool). Adding an electricity consumption size class variable shows whether the household consumes a lot of electricity or is below the median. The reports on electricity[116] and gas[117] consumption show how energy consumption varies by energy rating. The least energy efficient dwellings may use less energy because of an energy poverty problem.

The Netherlands: The Dutch Energy Poverty Monitor

The Dutch Ministry of Economic Affairs and Climate Policy commissioned Statistics Netherlands to develop a monitor providing insights into energy poverty among households at both national and regional levels. This initiative aligns with European obligations[118] to include energy poverty monitoring in climate action plans. In the development of the Energy Poverty Monitor[119], Statistics Netherlands collaborated with other experts in the field like the Organization for Applied Scientific Research to build upon pre-existing expertise (see the Organization's report). The developed Monitor, disseminated in January 2023, presents data from 2019 and 2020 and introduces four indicators (see Figure 6.2) to shed light on different aspects of energy poverty. These indicators encompass issues such as the impact of a home's energy efficiency on energy consumption and the interplay between high energy costs and insufficient income. Additionally, Statistics Netherlands provided a comprehensive microdata set, enabling external researchers to do in-depth research into energy poverty, using the same data as used for the monitor (the process is designed in such a way that the confidentiality of individual data is respected). Each year, the Monitor will be updated with recent data and, when relevant, new indicators and methodological refinements.

Figure 6.2.
Energy Poverty Monitor in the Netherlands

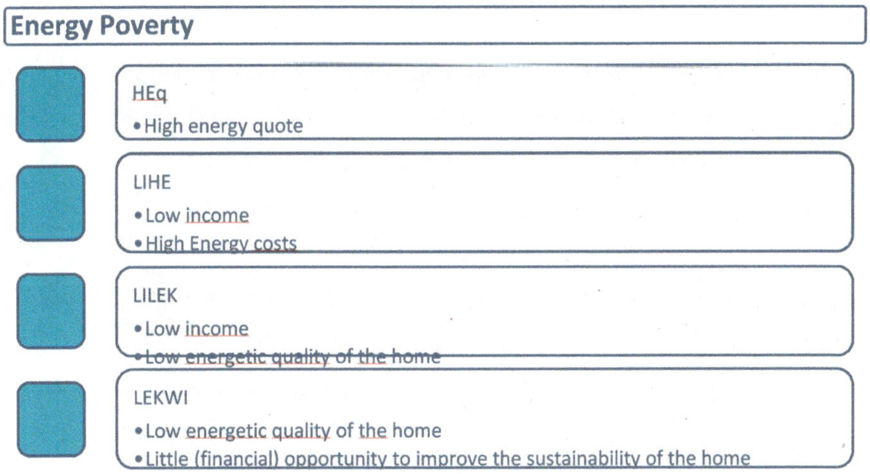

Source: Statistics Netherlands

Armenia: Identifying vulnerable groups

To achieve the Armenian mitigation goals the Government of Armenia adopted the Program of Energy Saving and Renewable Energy (2022-2030). Energy efficiency in the housing sector needs to be improved, which could

[115] https://www.cso.ie/en/statistics/social/domesticbuildingenergyratingsfromasocialperspective/
[116] https://www.cso.ie/en/statistics/energy/householdelectricityconsumptionbybuildingenergyratings/
[117] https://www.cso.ie/en/statistics/energy/householdgasconsumptionbybuildingenergyratings/
[118] https://energy.ec.europa.eu/topics/markets-and-consumers/energy-consumer-rights/energy-poverty_en
[119] https://www.cbs.nl/en-gb/background/2023/47/methodology-report-energy-poverty-monitor

impact the well-being of affected households. To ensure a just transition, i.e., a gender-responsive and socially oriented policies, the vulnerable groups were identified.[120] The following data sources were used: (1) Survey on Income and Living Conditions, (2) administrative data on socially vulnerable population, and (3) sociological surveys on vulnerability and gender at community level, including a case study on heating behaviour of female and male headed households. In general, one needs: (1) regular official statistics, (2) a further exploration of administrative data, and (3) extra surveys. The data used to identify socially vulnerable groups is generally at the national level, by urban and rural regions, and, in some cases, at the community level.

6.4 Conclusions and recommendations

491. NSOs can play a central role in monitoring a just transition in collaboration with policymakers, researchers, and socially vulnerable groups. Much of the data is already available in regular official statistics provided by NSOs, although the existence of such data may be unknown outside the statistical community. Similarly, further exploration of administrative data may reveal additional useful data. If necessary, additional surveys can be carried out by NSOs. To achieve proper (micro)data linking, NSOs can facilitate data sharing, in which the confidentiality of individual data is respected.

492. Based on this conclusion (and the preceding text), the following recommendations are made:

(a) Further work on developing harmonized indicators to measure progress towards a just transition is needed, for example on energy poverty or on skills needed for the green and decent jobs that will be available after the transition;

(b) To integrate a just transition perspective into NSOs' work on climate change-related data and statistics, an inventory of best practices is needed, including how to engage with stakeholders, how to incorporate social equity and inclusion considerations into data collection and analysis, and how to ensure that data is used effectively to support policy and decision-making;

(c) Given the importance of collaboration and engagement with stakeholders, there is a clear need for capacity building and data governance frameworks that enable producing insights relevant for just transition;

(d) NSOs can play a central role in developing methodologies to leverage and cross-analyse traditional household surveys, disaggregated socioeconomic data, and other (micro)data to identify information relevant to monitoring just transition (while respecting the confidentiality of individual data).

[120] See the paper: https://unece.org/statistics/documents/2023/08/working-documents/climate-change-related-gender-and-social and presentation: https://unece.org/statistics/documents/2023/08/presentations/climate-change-related-gender-and-social-vulnerability

Chapter 7 Informing the public

7.1 Introduction

493. The purpose of the chapter is to cover aspects of the role of NSOs in informing the public about climate change-related statistics and data.

494. The first section (section 7.2.1) describes the policy context for NSOs informing the public about climate change-related statistics, starting from the concept of action for climate empowerment. The second section (7.2.2) provides a schematic overview of user groups of climate change-related statistics and different user needs that the NSO can consider when publishing statistical products related to climate change. Results of a set of structured interviews with journalists conducted by the members of the Task Force are presented and analysed. Experiences from the journalists include which statistics are most interesting and which are the most difficult to access. The third section (7.3) on how the NSOs can contribute, covers the role of NSOs as well as examples of statistics, dissemination and organization in the NSOs. Conclusions are based on the results of the structured interviews as well as country examples collected by the Task Force.

7.2 Context and definitions

7.2.1 Policy context

495. The United Nations Framework Convention on Climate Change, the Paris Agreement and the 2030 Agenda for Sustainable Development all recognize that their goals cannot be achieved without involving all levels and sectors of government, the private sector, civil society and the public. Special efforts to educate, empower and engage all the stakeholders are needed to enable this public participation. This is expressed in Article 6 of the Convention and Article 12 of the Paris Agreement, presented in Box 7.1, and referred to as "Action for Climate Empowerment" (ACE).

Box 7.1.
Action for Climate Empowerment in the United Nations Framework Convention on Climate Change and the Paris Agreement

Article 6 of the Convention

In carrying out their commitments under Article 4, paragraph 1 (i), the Parties shall:

(a) Promote and facilitate at the national and, as appropriate, subregional and regional levels, and in accordance with national laws and regulations, and within their respective capacities:
  (i) The development and implementation of educational and public awareness programmes on climate change and its effects;
  (ii) Public access to information on climate change and its effects;
  (iii) Public participation in addressing climate change and its effects and developing adequate responses; and
  (iv) Training of scientific, technical and managerial personnel;

(b) Cooperate in and promote, at the international level, and, where appropriate, using existing bodies:
  (i) The development and exchange of educational and public awareness material on climate change and its effects; and
  (ii) The development and implementation of education and training programmes, including the strengthening of national institutions and the exchange or secondment of personnel to train experts in this field, in particular for developing countries.

Article 12 of the Paris Agreement

Parties shall cooperate in taking measures, as appropriate, to enhance climate change education, training, public awareness, public participation and public access to information, recognizing the importance of these steps with respect to enhancing actions under this Agreement.

496. ACE advocates that all citizens must be empowered to take individual climate action. To do so, they need to be sufficiently informed. Therefore, improving public outreach, education and awareness of climate issues will be crucial for the transition to a low-emission, climate-resilient world.

497. At the 11th Conference of the Parties in Doha, the governments adopted the Doha Work Programme, which provides guidance on how countries should implement ACE (UNFCCC Subsidiary Body for Implementation, 2012). The Doha Work Programme recommends a long-term, strategic and country-driven approach, including integrating relevant activities into existing climate change adaptation and mitigation programmes and strategies. Each country has also been requested to nominate a National Focal Point to coordinate these efforts.[121]

498. Based on the UN Framework Convention and the Paris Agreement, ACE is considered to have six main aspects:

(a) Education;

(b) Training;

(c) Public awareness;

(d) Public access to information;

(e) Public participation;

(f) International cooperation.

499. The objectives of each aspect are presented in Table 7.1.

Table 7.1
Article 6 elements - scope and objectives

Scope	Objectives	
Education	Change habits in the long-term	Foster a better understanding of, and ability, to address climate change and its effects
Training	Develop practical skills	
Public awareness	Reach people of all ages and walks of life	Promotes community engagement, creativity and knowledge in finding climate change solutions
Public access to information	Make information freely available	
Public participation	Involve all stakeholders in decision making and implementation	Engage all stakeholders in debate and partnership to respond collectively to climate change
International cooperation	Strengthen cooperation, joint efforts and knowledge exchange	

***Source*: UNFCCC and UNESCO (2016)**

500. In the context of official statistics, the two key ACE elements are **public awareness** and **public access to information**, which both aim to share information and knowledge with the public to eventually support and enable public participation in climate action.[122]

[121] An up-to-date list of national ACE focal points is available at the UNFCCC web page: https://unfccc.int/topics/education-youth/national-ace-focal-points

[122] https://unfccc.int/topics/education-and-outreach/workstreams/public-participation

501. Public awareness initiatives aim to improve understanding of climate change, which is expected to impact attitudes, behaviours and choices. The media is considered one of the most important communication methods and will include television, radio, print and social media. These can be used to spread and disseminate information on the causes and effects of climate change as well as meaningful and practical solutions (UNFCCC Subsidiary Body for Implementation, 2012).

Did you know?

Publishing official environmental statistics is one of the examples of implementation of Action for Climate Empowerment in the area of Public Access to Information reported by the countries in their national contributions.

502. Actions aimed at improving **public access to information** aim to make information freely available, transparent and accessible. Improving access to information involves not only disseminating information but also supporting or creating mechanisms that help share the knowledge across sectors and communities, and between decision-makers at all levels, practitioners and citizens. Information, including data and statistics, is needed to understand, address and respond to climate change, including information on climate change initiatives, policies and results of actions (UNESCO and UNFCCC, 2016).

503. To contribute to the key ACE elements, NSOs need to communicate and disseminate data and statistics to a number of different audiences and in turn, be responsive to user needs. The next section provides an overview of different users, together with their needs. General examples are provided alongside.

7.2.2 Audiences and users of climate-change related statistics

504. Users of climate change-related statistics are many, and they differ in their data needs and knowledge. Some users of climate statistics have a deep understanding of the topic and are used to handling data and statistics, while others have little or no prior knowledge and are not experienced in reading raw data or complicated tables and visualizations. Some may even consider the topic of climate change statistics too complex or overwhelming.

505. Increasing public interest in a broad range of climate change-related issues, including emissions and mitigation / net zero policies, impacts on communities and the environment, the increasing need and benefit of adaptation and interconnected links with other topics such as the economy, health, transport and policy, brings new and emerging user groups of climate change-related statistics. In turn, this increases the need for NSOs to map and understand users and their needs.

7.2.2.1 Who are the public?

506. In this chapter, we focus on the public as key recipients, users and demanders of climate change-related information. Here, the public comprises primarily private individuals, who may have many different information needs. They may be interested in climate data for their own information or to participate in conversations with friends or at work, or to be involved in politics or climate activism. They may also need climate change data in their professional occupation, e.g. for preparing policy or legislation, providing training, or giving lectures or interviews as an expert, for strategic planning or for developing data-driven apps etc.

507. The mainstream media along with social media, NGOs and civil society all play a role in making information and data available to the public. For the purpose of this chapter, public need does not include data access and transfer between NSOs and academia, but it does include activities of individual researchers when undertaken as private individuals (perhaps with a specific or professional need for climate change-related data).

7.2.2.2 User groups and their needs

508. Recipients and users of climate change-related data and statistics can be categorized into four user groups. These groups are defined, for the purpose of this report, by how users come into contact with climate data and statistics (for example, passively receiving or actively searching) as well as their prior understanding and knowledge of climate change and data analysis. Some examples are given for the constitution of each of the user groups, even though some groups of individuals (such as journalists) may be present in more than one user group.

Table 7.2.
User groups for climate change related statistics

Generally interested citizens	Specifically interested	Professionals and policymakers	Analysts
• Passively receiving or actively searching for information • Encounter data via press or social media, blogs, publications or internet searches • Infrequent visitor to statistics-related websites • Without special or professional interest in climate change data	• Actively searching for information • Looking for unbiased facts about topical issues • E.g. general journalists, teachers, students or consultants	• Actively searching for information • Prefer official statistics/ NSO as data source • Frequent users of statistics for benchmarking and comparison • Able to withdraw, customize and present data • May have own agenda or own media, data gives them credibility • E.g. climate journalists, employees in ministries, agencies, private companies and NGOs, climate activists.	• Actively searching for information • Require official statistics/ NSO as data source • Data experts, data miners • Able to analyse and document data • Less interested in written publications • E.g. data journalists, researchers, climate experts, data-driven app developers

Source: Authors.

509. User groups may have very different needs, and NSOs should consider each user group when producing and disseminating data and statistics. A schematic illustration of the typical needs for climate change-related statistic products is shown below.

Table 7.3.
Needs for climate-change related statistics products by user group

	Generally interested	Specifically interested	Professionals and policymakers	Expert analysts
Overview content Interactive maps, infographics, graphs Key figures, dashboards Simple wording, engaging visuals and titles	■	■	■	■
Commented statistics Press releases, reports, analysis Thematic webpages		■	■	■
Statistics Tables Documentation of statistics			■	■
Data, including microdata Tailored and detailed data Anonymized micro data Geospatially enabled data				■

Source: Authors.

510. **Overview content** should address the needs of all user groups and should be very easy for the public to access. Examples include interactive graphs and maps, climate figures and infographics. Most content will be suitable for publication on social media and be accessible to a non-technical audience. 'Generally interested' users will likely be satisfied with overview content, while 'specifically interested' users, 'professionals and policymakers' and 'expert analysts' may use this as a starting point for researching further. References to other relevant data or information sources (to underlying datasets or links to policy, for example) may also be part of overview content.

511. **Commented statistics** will be slightly more technical than overview content and might include press releases, explained statistics, shorter articles, reports or thematic web pages. These products may be too specific or in-depth for 'generally interested' users but are expected to suit the needs of users that are 'specifically interested' in identifying or finding climate change topics. 'Professionals and policymakers' as well as 'expert analysts' may use commented statistics as a starting point for researching further or for more detail.

512. **Statistics** will comprise more technical tables and in-depth articles or reports. These might be available from NSOs on different climate topics and might include documentation of methodology, sources and data treatment. This format will likely require a significant level of thematic or analytical knowledge and will likely be unsuitable for 'generally interested' and 'specifically interested' users but should satisfy the needs of 'professionals and policymakers'. Analysts may use these statistics for creating novel visualizations, analyses or as background or supporting evidence in reports or articles. They may also be used as a starting point for further work, such as mapping additional data availability to structure micro-level data analysis.

513. **Data, including microdata,** (under special conditions) may be provided by NSOs to certain data users, by data request or via specific user searchers. This highly granular data will be too complicated or technical for 'generally interested' or 'specifically interested' users as well as most 'professionals and policymakers', although some users from this group may find it useful. However, 'expert analysts' should find this data useful for individual research or analysis.

7.2.2.3 New and emerging users of climate change related statistics

514. As analytical methods become more advanced, a number of new and emerging user groups for climate change-related statistics emerge. These might include developers of apps or websites focused on interactive decision-making tools such as carbon footprint or carbon offsetting calculators or climate-friendly travel decision tools, for example. These tools may rely on emissions or other climate related data to underpin their functionality.

515. Individuals involved in generating social media content, like bloggers, vloggers, influencers or climate activists, may also have an increasing demand for climate change-related information, data and statistics.

516. In addition to these, there is an increasing demand to understand the links and connections between climate change and topics such as finance and business, health, public behaviour, transport and the environment, from which there may be considerable demand and interest across a range of climate change topics. The finance sector is also a user group with increasing demand, especially for timely up-to-date data on climate change related aspects.

7.2.2.4 Experiences on the interest and information needs of the public

517. The media are an important source of information for the public. Journalists are a key group responsible for obtaining and disseminating information, data and statistics to the public. They should therefore have a good understanding of what topics are of current public interest and the best ways to effectively reach their audience.

518. During 2022, members of the Task Force interviewed journalists from across the UNECE region, using a structured interview technique and a set of pre-prepared questions. This enabled a systematic collation of answers, which were then compared and analysed.

519. Interviewed journalists comprised more and less technically competent individuals, covering a range of climate-related topics, and included those working on traditional story-focused narratives as well as those that prefer to create novel graphics, visualizations and analysis. Some of these journalists have specialized data or GIS analysis skills and often work on stories with data as an essential element.

520. A summary of the interviews and a set of key findings on how NSOs can help to better inform the public are provided below. Individual anonymized replies are presented in Annex 5.

Climate-related data of most interest to the public

521. Journalists working on climate change stories often cover a broad set of interconnected issues spanning environmental, social and economic topics. These can include the wider environment, energy, agriculture, transport, business and politics.

522. Stories can cover the underlying science of climate change as well as its impacts on people and communities that may focus on small-scale local action, or on global long-term trends and comparisons.

523. The following are some examples of stories:

(a) Extreme weather and its impacts - for example, rainfall, drought, rising temperatures like the European heatwave experienced during summer 2022, or climate projections;

(b) Civil society mobilization;

(c) Local interest, for example, how cities can adapt, local impacts on communities, local action;

(d) Agricultural topics, natural resources and wider environmental issues;

(e) Geo-politics;

(f) Global climate science stories;

(g) Energy;

(h) Transport;

(i) Science.

Sources most used by climate-journalists

524. Journalists use a **variety of data sources**, depending on the story. This includes official and hydrometeorological statistics as well as press releases and reports from both home and international sources.

525. Time, resources and technical skills often dictate which sources are used. Journalists with more advanced technical skills tend to prefer creating unique visualizations and use sources that provide **access to underlying datasets**. Less technical journalists, or those on tight timeframes, often prefer to use **ready-made reports and analyses**. There can be difficulty understanding which source to use and when.

526. Public information services that provide the **ability to search and query** content can be very helpful in supporting journalists to locate appropriate information without the need for in-depth research and data mining that require significant resources and technical skills.

527. Some journalists mentioned using **social media** to look for data, which has some advantages, particularly in terms of timeliness and responsiveness. At the same time, this comes with a high risk of coming across exaggerated or false information. However, this approach can be particularly useful for identifying, corroborating and supplementing other data sources.

528. Official sources of information are recognized as **critical** to good journalism.

529. For those working on global topics, the language and format of the report, data or statistics are of primary concern. Resources that are in **English** and are easily accessible will be accessed and used more often.

Which climate-related data are difficult to access

530. Most journalists report that **physical metrics,** such as the weather, emissions or energy, are generally well available at the country level.

531. More difficult to access are social, environmental and economic metrics, such as those that might inform on climate change impacts or adaptations, or on the interactions between climate change and other areas (such as finance, society, health, nature, etc.). The following are some specific examples identified by journalists:

(a) **Climate adaptation;**

(b) **Businesses** and **finance**:

(i) Financial climate change aid sent to other countries;

(ii) Emissions data from large individual companies;

(iii) Interactions between climate risks and economic damage;

(c) **The impact of climate change on both health and nature;**

(d) Consumer **carbon footprint** metrics.

532. Aside from topic-specific issues, journalists reported four difficulties: (1) coverage, (2) timeliness, (3) granularity, and (4) accessibility.

533. Considering **coverage**, issues with different scales, geographies and lack of coherence as well as lack of availability, particularly for parts of Africa, Asia and Latin America and others outside of the OECD were mentioned.

534. **Insufficient timeliness** concerned both periodic, e.g. annual publications based on historical data and up-to-the-minute data. Timeliness of data releases has different degrees of importance across journalists, to some journalists timeliness is not a priority while it is very important to others.

535. **Granularity** issues stem from inconsistencies in the availability of data collected or provided at a sub-national level. This refers to a lack of consistency in what data is collected at this level within a country, the scale at which it is collected and the statistics that are provided, analysed and shared by local or regional data providers. Journalists often want more local-level data to inform on local stories but face difficulties in finding data at the right scale for the right place.

536. **Accessibility** issues are two-fold and often contradictory. Some journalists want more consistently available data in open formats to allow easy and speedy access for novel analyses and visualizations. Others do not have the skills, resources or time for this and want more accessible reports with ready-made analysis and visualizations that can easily contribute to, supplement and enhance narratives.

537. For less technical journalists, there can be difficulties navigating complicated scientific reports where specific data, analysis and conclusions may be difficult to find.

538. Comments include:

(a) NSO service desks are useful facilities to support journalists in contacting the NSO and querying data, data products and reports;

(b) The use of multiple and conflicting reference periods (e.g. in the context of temperature increase) is confusing. Standardized and fewer reference periods would be helpful.

539. In summary, less technical journalists need triggers like press releases, preferably based on integrated reports. Particularly helpful are:

(a) Fact sheets and dashboards where the most relevant data can be searched and selected;

(b) Pre-releases under embargo in order to publish the moment the data becomes officially available.

540. Technical journalists need reliable and independent data which are easily available, for example:

(a) Geospatial data in simple, usable standard formats (for example, shape files, JSON, vector product format, etc.);

(b) Free, open-source data (for example, available through GitHub or Tableau, in CSV format, etc.).

541. Please note that these comments reflect the experiences of journalists only within the UNECE region and we cannot comment on data availability more widely.

542. On the supply side, for NSOs there may be different areas in climate change-related statistics, where they are not able to provide the answers to questions from users. Some of these challenges are related to granularity of GHG emission data and statistics:

(a) At the regional or provincial level;

(b) For certain industries, special activities (e.g. heating residential buildings) products or production processes. For small, medium and large enterprise categories;

(c) By pollutant;

(d) By income level for household emissions or footprints.

7.3 How NSOs can contribute

7.3.1 The role of NSOs in informing the public

543. The first Fundamental Principle of Official Statistics states that "Official statistics provide an indispensable element in the information system of a democratic society, serving the government, the economy and the public with data about the economic, demographic, social and environmental situation. To this end, official statistics that meet the test of practical utility are to be compiled and made available on an impartial basis by official statistical agencies to honour citizens' entitlement to public information" (United Nations, 2014).

544. Informing the public is at the core of the mandate of NSOs. This is achieved by compiling and making available data on relevant topics from the economic, demographic, social and environmental domains, in an impartial way, following standards, using proper sources, and accompanied by metadata. With these principles, the role of NSOs in informing the public is crucial as they have been widely considered trusted data providers. This role is even more important in the present time when data from unverified sources are abundant, and information and disinformation are spreading at a rapid pace. NSOs also contribute to a wider social goal of transparency, accountability, and building a relationship between the state and citizens by ensuring the availability of trusted statistics to the public. They enforce their role as a key component of the information system of a democratic society by providing reliable statistics on issues relevant to the public. To determine what is relevant, it is particularly important that NSOs understand the users and potential users of statistics and their information needs.

545. This principle emphasizes one important aspect of how official statistics serve the public - not only by being compiled but also by being made available to the citizens. NSOs have decades-long expertise in disseminating their products in an impartial, transparent and accessible way, and have been putting a lot of focus on how to communicate statistics effectively. This expertise can be very valuable in the climate change domain, where there are multiple data compilers, often outside of the statistical system, who may not have had an equally strong focus on making their data available and

easily accessible to the public. NSOs can play a role in communicating the climate topic in a purely factual manner, especially in potentially disputed areas where there may be national or local political controversy in climate-related issues. NSOs can carefully consider the national aspects to strike the balance to remain impartial while presenting information that has high policy relevance.

546. In practice, NSOs first and foremost contribute to informing the public by publishing relevant and high-quality statistics. This includes GHG emissions data, but also other types of climate change-related statistics, as is described in the next section.

7.3.2 Producing climate change-related statistics of public interest

547. NSOs or other producers of statistics compile and make available statistics on GHG emissions, which can be compiled using the SEEA framework emissions accounts. Even in cases when NSOs are not the main data compilers, NSOs can provide guidance to the users when it comes to the different methods of compiling emissions data, IPCC and SEEA-type statistics or production-based vs. consumption-based emissions approaches. Examples are also provided in Chapter 4. An example of explaining measurement perspectives on air emissions is provided by Eurostat.[123]

548. It can be useful to consider the user groups and their needs so that an easy-to-access text box describing the most important methodological concerns is available, potentially with a technical description available for more technical users. Metadata or documentation of statistics can be used to provide more information, which can help the user assess the fitness of use for different purposes. An example is provided by Statistics Canada, where the user can choose upfront which method is preferred, after being presented with an overview of the National Inventory Report approach and the Physical Flow Account (PFA) approach. Several tables are provided to help the user compare the two approaches in terms of their scope, reporting framework, classification system, and principles, as well as which pollutants are included.

549. NSOs may also collect, make available and use in the production of additional statistics, alternative emissions data and statistics from other national agencies. This might include data from the national emission inventories, if these are produced outside of the NSO, or energy statistics from energy agencies.

550. Aside from emissions data, NSOs may provide other climate related statistics. Journalists provided some examples of topics they also had an interest in:

(a) Foreign climate-related aid;

(b) Climate-related health expenditure;

(c) Figures and values that show how different aspects of climate change relate to each other;

(d) Waste management;

(e) Low-carbon energy consumption;

(f) Connecting micro data on businesses with their GHG emissions.

7.3.2.1 Geospatial data

551. The demand for local and detailed climate change-related statistics is linked with the use of geospatial data in geospatial data for climate change-related statistics, which adds location information (coordinates, municipality codes, regional information etc.) to the statistics. This demand includes both highly granular data for specific locations and more general demand for data at subnational level.

552. One example from Denmark is a recent project on grouped municipal emission accounts averages for urban and rural areas.

[123] https://ec.europa.eu/eurostat/web/environment/information-data/emissions-greenhouse-gases-air-pollutants

553. More technical journalists have expressed a need for geospatial datasets to show and explain climate-related data via novel spatial visualizations or infographics.

554. There are multiple issues to consider related to how NSOs might meet the demands for geospatial statistics in the climate change domain. Some are below:

(a) Source data may not include geospatial data;

(b) Whether the approach should be top-down (splitting the national total down to smaller local entities) or bottom-up (adding up the emissions for companies at a sub-national level to a national total) and what to do with discrepancies between methods;

(c) Priority and resource allocation within the NSO to enable and support the development of high-quality geospatial data.

7.3.3 Dissemination and communication of climate-related statistics

555. Different users have different needs and are interested in different statistical products. Below is a list of country examples as well as journalists' examples of good data and statistical products:

(a) Europe 1 degree warmer, available at https://www.onedegreewarmer.eu/;

(b) Thematic website on climate change-related statistics in Denmark, available [in Danish] at https://www.dst.dk/klima;

(c) Thematic website on climate change related statistics from the statistical office of Switzerland available [in French and German] at https://www.bfs.admin.ch/bfs/fr/home/statistiques/espace-environnement/indicateurs-lies-au-climat.html;

(d) Dedicated government climate web page from Mexico at https://gaia.inegi.org.mx/sicc/;

(e) United States Energy Information Administration and "Today in Energy" blog, available at https://www.eia.gov/todayinenergy/;

(f) Disseminating geospatial data and World Resources Institute, available at https://www.wri.org/data;

(g) United States National Oceanic and Atmospheric Administration, available at https://www.noaa.gov/climate;

(h) UK Climate Change Committee for less complicated and accessible datasets, https://www.theccc.org.uk/;

(i) CE Delft's practice of disseminating figures on individual companies, https://cedelft.eu/.

556. A general challenge facing climate analysts is to go from the global to the individual perspective. An aspect that NSOs may consider is enabling links between the statistics and effects on the individual. E.g. effects related to private transport, income and electricity consumption may increase the relevance of climate change-related statistics for the public. Production of studies at a small geographical scale (e.g. city or regional level) can be useful to link the local to the global perspective.

7.3.3.1 The role of NSO as an active data provider, proactively reaching out to public

557. Journalists often use data provided by NSOs and consider NSOs as playing important role in providing and disseminating data and statistics - at least as important as government ministries and hydrometeorological services in most countries.

558. Official sources of climate change-related statistics, including NSOs, government ministries and hydrometeorological organizations, are primary sources for many journalists, but the importance and impact of each source in communicating and disseminating to the public varies across countries.

559. NSOs may provide climate-related data to the public in various ways, including both static and interactive content covering statistical yearbooks, bulletins, articles, websites, portals, dashboards, social media, tables, graphics, geospatial data and mapping tools, and downloadable files etc.

560. In addition, NSOs can proactively engage in outreach activities. Many of these may be focused on schools and education institutes to support teaching climate change topics and to engage students around climate change and climate change-related statistics. One example of this from Denmark is a project designing a statistics-based interactive learning platform for schools.

561. NSOs may also, in cooperation with other national organizations, host conferences or thematic meetings on the topic of climate change, statistics and data, inviting a debate for the public as well as experts in the climate field. Two such examples from France from 2023 can be mentioned:

(a) By the French Economic, Social and Environmental Council: https://www.lecese.fr/actualites/retour-sur-les-rencontres-du-cese-climat-comment-financer-la-transition (in French);

(b) By the French National Council on Statistical Information: https://www.cnis.fr/evenements/colloque-mesurer-la-transition-ecologique (In French).

562. Statistics Canada provides users with an interactive data visualization tool.[124] The tool allows users to visually compare industries across a time series via an online platform. The user can decide which industries and time period is of interest. The tool is aimed at the public, students and teachers as well as the media and industry groups looking to compare emissions.

7.3.3.2 NSOs actively using social media or other platforms

563. Use of social media to disseminate statistics is not as widespread as traditional methods but is on the rise. The UK Office for National Statistics, for example, has just started conveying key statistical points via Instagram for the first time, having used X, formerly known as Twitter, regularly for several years.

564. Journalists are aware of social media as a potential dissemination channel for public awareness on environmental issues but also as a means of delivering public scrutiny. However, this comes with a known risk of encountering exaggerated or false information, so it should be used with caution.

565. Increasing the use of social media (e.g. Facebook and LinkedIn) to disseminate climate change-related statistics would be beneficial in increasing outreach and engagement with sections of the public that may not already engage with traditional dissemination products.

566. Some challenges may come with this approach. NSOs will need to consider that resources for timely engagement with social media content and users will be crucial to making effective use of this format. NSOs may also need to consider big data management tools. These needs will likely develop further if and when the use of this format increases.

567. NSOs can provide links or guidance for users to find other sources, e.g. hydrometeorological data and climate-related legislation, to help users find more relevant data - or become more informed on climate-related topics.

7.3.3.3 Machine readability, user support and engaging with the public

568. Generally, NSOs disseminate many indicators, and journalists often find it difficult to know and find the most appropriate sources. On the other hand, in many cases, users need more detailed information for specific projects or stories.

[124] https://www150.statcan.gc.ca/n1/pub/71-607-x/71-607-x2020008-eng.htm

569. Less technical journalists, or those with tight timeframes or limited resources, often prefer to avoid undertaking lengthy or complicated research and instead opt for more readily available reports or use services that can assist in finding appropriate resources effectively and efficiently.

570. More technical journalists generally prefer to access data that is machine-readable and easy to access and extract.

7.3.4 How to work with the press office/communication experts within the NSO and mapping and evaluating user needs

571. NSOs have dedicated communications departments that often include strategic communications, external affairs and media relations offices. Focusing on different elements of communications, dedicated teams guide and coordinate communication activities; collate and coordinate stakeholder insight, impact and engagement; and promote and disseminate work through external media channels, respectively. Collaboration between domain experts and communications experts can be particularly helpful in such a new and complex area as climate change.

572. The communications department and the statistical experts at Statistics Denmark, for example, have a short weekly meeting to discuss potential publications or ideas for posts on social media (the meetings cover all areas of statistics, not only climate-related data). The stories can be triggered by recent events in the media, international climate meetings or new ways of integrating data from different statistical domains.

573. NSO statistical and data publications are often combined with media or press releases to publicize content and inform journalists of key points. Media relations offices can therefore be a critical route to highlighting new data sources and statistics and disseminating information to the public.

574. User engagement plays a key role for NSOs and might include thematic (climate) focus groups, analysis of website usage (via Google Analytics, for example) and feedback surveys (such as hotjar). They can be a good way for NSOs to map the demand for information and evaluate if the chosen approach/design of websites, publications and statistical products effectively fulfils user needs.

7.4 Conclusions and recommendations

575. The role of NSOs in informing the public about statistics and data relating to climate change ties in with the principles of ACE as well as the Fundamental Principles of Official Statistics. There are several user groups for climate change-related statistics, and preferences differ in terms of communication products, channels and dissemination of data.

576. The public is interested in both in-depth stories and insight-driven graphics on a wider range of climate change topics spanning environmental, social and economic issues. However, there are significant gaps in data availability, particularly on the impacts of climate change, its interactions with health, the economy and nature, and of adaptation. There are also broader difficulties with the level of granularity and accessibility of data.

577. Country examples provide insights into how the NSOs can work to disseminate and communicate climate change-related statistics, actively reach out to the public, use social media and other channels, work on machine readability and searchability of statistics and coordination between statistical experts and communications staff within the NSO.

578. Recommendations:

(a) To ensure and contribute to public awareness and public access to information on climate, the NSOs can map and regularly evaluate user needs and user experience, specifically with climate change-related statistics. Tools like focus groups, analysis of website use and feedback surveys can be used;

(b) New and emerging user groups and technologies underline the need for NSOs to communicate widely, proactively as well as improve machine readability and searchability of statistical products and websites;

(c) NSOs could assist users by collecting links or resources to relevant sources such as national legislation, temperature and weather data or international comparisons. This can be done through thematic websites or dashboards for climate-related statistics and indicators;

(d) Data content on platforms such as thematic pages and dashboards can be presented according to the structure "drivers- emissions- impacts- mitigation-adaptation". International comparisons and information available in English increase the relevance of such channels for some users;

(e) NSOs can provide user guidance on the different methods/approaches for emissions compilation, ideally both for generally interested users in a short format (e.g., key figures or short fact box) as well as for expert analysts via e.g. technical notes and documentation.

Chapter 8 Climate-related costs, transfers and financial flows

8.1 Introduction

579. Climate finance, which is, so far, a concept without an agreed international definition, is an enabler of climate action described in the previous chapters - for climate change mitigation, adaptation and just transition. Monitoring the quantity of financial flows is a means to assess the effort made towards the transition to a low-carbon and climate-resilient economy.

580. At the same time, the financial dimension of climate change is broader. Investors consider companies´ stance vis-a-vis climate change and sustainability objectives (energy use, waste management, pollution) - which has led to increase in environmental, social and governance (ESG) reporting. Climate change has also become a key concern for international regulators and a more important factor for the stability of national and international financial institutions, who are exposed to climate related risks, associated with financial stability concerns.

581. As the topic is broad and complex, and the official statistical community's contribution is still in fairly early stages, this chapter aims to outline the main issues relevant for statisticians, informs about initiatives already in place and initiates the discussion about the potential role for NSOs in this landscape, complementing the efforts of other actors in the public and private sector.

8.2 Policy context, definitions and data landscape

8.2.1 Defining, tracking and measuring climate finance

582. Making financial flows consistent with a pathway towards low GHG emissions and climate-resilient development, is one of the goals of the Paris Agreement (Article 2.1c) - next to the mitigation and adaptation goals. The Convention, the Kyoto Protocol and the Paris Agreement try to mobilize enough financial resources and help countries adapt to climate change effects, including through mutual support. Specifically, article 9 of the Paris Agreement requires developed country Parties to provide resources to assist developing country Parties with respect to both mitigation and adaptation. Reporting on the support provided and received is part of the Enhanced Transparency Framework. Assessing progress in provision and mobilization of support is also part of the global stocktake under the Agreement.

583. Under UNFCCC, a financial mechanism has been established for developed countries to provide financial resources to developing countries and hence "facilitate the provision of climate finance". This financial mechanism serves the Convention, the Kyoto Protocol and the Paris Agreement. Its operating entities have been the Global Environment Facility (GEF) and the Green Climate Fund (GCF) together with the Special Climate Change Fund (SCCF) and the Least Developed Countries Fund (LDCF) managed by the GEF–and the Adaptation Fund (AF). To assist the COP in exercising its functions in relation to the financial mechanism, the Standing Committee on Finance (SCF) was established in 2010. Since 2015 it has also been serving the Paris Agreement. The functions in which the Committee is assisting and supporting the COP are:

(a) Improving coherence and coordination in the delivery of climate change financing;

(b) Rationalization of the financial mechanism of the UNFCCC;

(c) Mobilization of financial resources for climate financing; and

(d) Measurement, reporting and verification of support provided to developing country Parties.

584. One of the key SCF activities supporting measurement, reporting and verification of support provided to developing country Parties (function (d) above) is preparation of a Biennial Assessment

and Overview of Climate Finance Flows (BA)[125]. The assessment draws on the available sources of information and including information on the geographical and thematic balance of flows.

585. The SCF has also been working on strengthening methodologies for reporting climate finance and operational definitions of climate finance, including private finance mobilized by public interventions, to assess how adaptation and mitigation needs can most effectively be met by climate finance.

Box 8.1
Long-term finance and the USD 100 Billion Goal

"At the 15th Conference of Parties (COP15) of the UNFCCC in Copenhagen in 2009, developed countries committed to a collective goal of mobilizing USD 100 billion per year by 2020 for climate action in developing countries, in the context of meaningful mitigation actions and transparency on implementation. The goal was formalized at COP16 in Cancun, and at COP21 in Paris, it was reiterated and extended to 2025.

At the request of donor countries, the OECD has been tracking progress towards this goal since 2015. It produces regular analyses of progress made, based on a robust accounting framework that is consistent with the COP24 outcome agreed by all Parties to the Paris Agreement on funding sources and financial instruments." (https://www.oecd.org/climate-change/finance-usd-100-billion-goal/)

"In Paris, (...), the COP extended the goal of mobilizing jointly USD 100 billion through 2025 and decided that prior to 2025 the Conference of the Parties serving as the meeting of the Parties to the Paris Agreement (CMA), shall set a new collective quantified goal from a floor of USD 100 billion per year, taking into account the needs and priorities of developing countries." Decision 1/CP.21, para. 53. https://unstats.un.org/sdgs/tierIII-indicators/files/13.a.1_Background.pdf

New collective quantified goal is expected to be adopted at COP29.

586. Although "climate finance" is an often-used term, it currently lacks an internationally agreed definition and is used differently in different contexts by different stakeholders. The current operational definition of climate finance used by the SCF is:

> *"Climate finance aims at reducing emissions, and enhancing sinks of greenhouse gases and aims at reducing vulnerability of, and maintaining and increasing the resilience of, human and ecological systems to negative climate change impacts."*

587. This definition was identified in the 2014 Biennial Assessment and Overview of Climate Finance Flows Report of the SCF, as a convergence of the various definitions in use.

588. The glossary entry for climate finance of the Sixth Assessment Report of the IPCC reads:

> *"There is no agreed definition of climate finance. The term 'climate finance' is applied to the financial resources devoted to addressing climate change by all public and private actors from global to local scales, including international financial flows to developing countries to assist them in addressing climate change. Climate finance aims to reduce net greenhouse gas emissions and/or to enhance adaptation and increase resilience to the impacts of current and projected climate change. Finance can come from private and public sources, channelled by various intermediaries, and is delivered by a range of instruments, including grants, concessional and non-concessional debt, and internal budget reallocations."*

589. Since the 3rd Biennial Assessment, the SCF has been working on definitions of climate finance and has found that the understanding of what climate finance encompasses varies including in terms of:

(a) Which sectors and activities are covered;

(b) Range of financial instruments included;

[125] https://unfccc.int/topics/climate-finance/workstreams/transparency-of-support-ex-post/biennial-assessment-and-overview-of-climate-finance-flows-background

(c) Which tracking and reporting processes apply; and

(d) Different perspectives on what definitions of climate finance should include and the detail in which associated concepts should be defined.

590. Building on that work and based on the submissions of views from Parties and non-Party stakeholders, in 2023, the Standing Committee on Finance published a report on clustering the different definitions (UNFCCC Standing Committee on Finance, 2023). It noted that the analysed definitions of climate finance are often developed for different purposes and use cases, their scope varies from domestic to international in context, the level of detail from high-level description to detailed classification of specific climate-related activities, and from focusing solely on what makes finance flows climate related to covering financing types or accounting-related boundaries of finance flows. The report lists all the clustering elements and decision points within each element to develop and apply a definition of climate finance suitable to the purpose of the definition.

Figure 8.1.
Summary of climate finance flows in 2019-2020, estimated in line with the operational definition of climate finance adopted in the 1st BA in 2014.

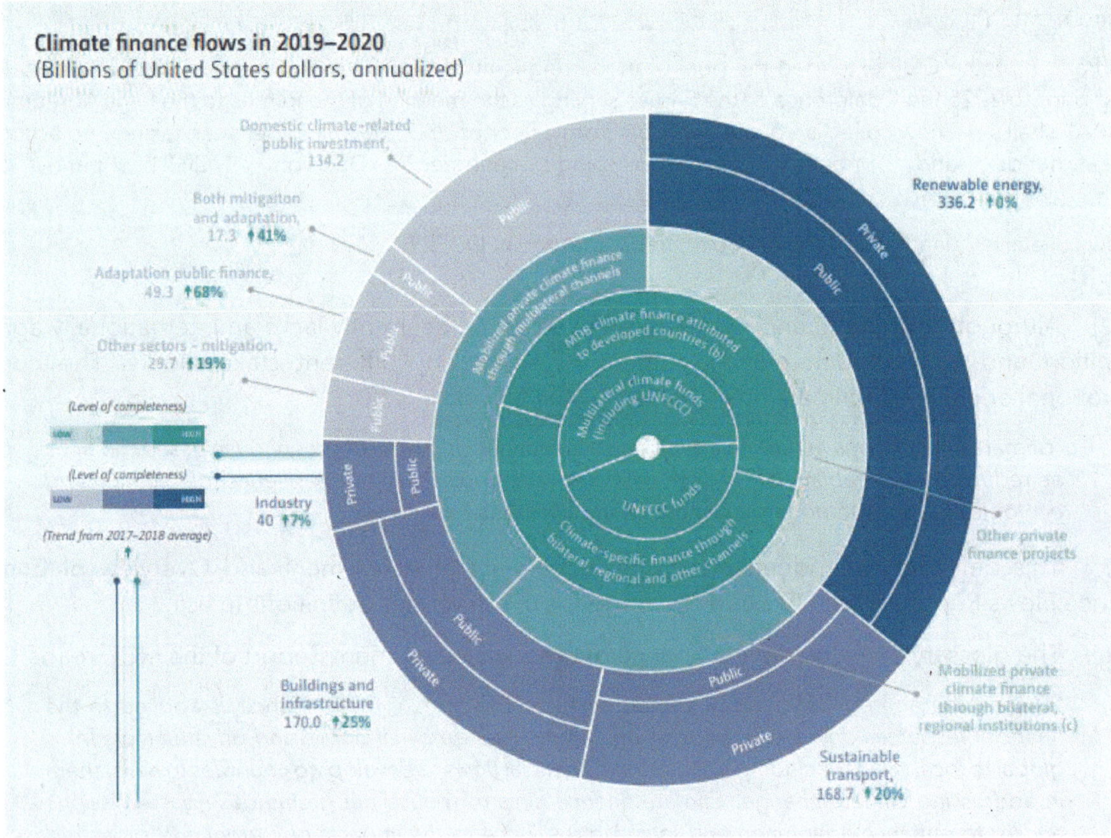

Source **UNFCCC Standing Committee on Finance (2022).**

591. In the 5th BA (UNFCCC Standing Committee on Finance (2022)), the SCF presented an assessment of climate finance flows in years 2019-2022 (see Figure 8.1 for a high-level overview). The Assessment identified the following methodological challenges and limitations in collecting and aggregating data on climate finance:

(a) In compiling the estimates, efforts were made to ensure that they are based on activities that are in line with the operational definition of climate finance adopted in the first BA in 2014 and to avoid double counting. Challenges were encountered in aggregating and analysing information from diverse sources with varying degrees of transparency.

(b) Regarding global climate finance, filling data gaps remain challenging, particularly on private finance for adaptation activities and for mitigation activities in the agriculture, forestry and other land use (AFOLU), waste, and water and sanitation sectors. Methodologies for calculating climate finance based on total cost or incremental cost produce different estimates by activity, which potentially leads to limitations regarding the completeness of data and any interpretation of the relative shares of global climate finance going to different themes or sectors. Energy efficiency estimates do not include data broken down by public or private actors financial instruments, or at country level. Some data sources, such as those for renewable energy, provide activity-level data but may make country- and technology-level assumptions on finance flows to fill data gaps.

(c) Regarding domestic climate finance, although more countries are developing climate finance reporting systems, time lags in implementation mean data are underreported for 2019–2020. Amounts in relation to public expenditure may refer to ex ante budget allocations or ex post actual expenditures; the climate relevance of activities reported may refer to weighted criteria per activity or to positive activity lists.

(d) Data on international climate finance flows are compiled using various methodologies and have varying interpretations:

(i) Flows from developed to developing countries - covering finance provided, mobilized and received - include a mix of data based on disbursements to projects and recipients in the given year or on financial commitments made in the reporting year to activities that may be implemented over several years.

(ii) Information on South-South cooperation in climate finance flows remains relatively underreported. The classification of data such as by geographical region or by granularity is not uniform across data sources.

(iii) As for previous BAs, no aggregation of data from different sources for finance flows from developed countries to developing countries was carried out owing to these challenges and limitations.

592. **The 5th BA also underlines variety of actors undertaking reporting on climate finance, with different purposes and processes.** The stakeholders include "providers of raw data: both public and commercial data providers, aggregators of data from various sources, publishers of climate finance estimates and Parties themselves, which report on climate finance support provided, mobilized and received (see Figure 8.2). Some actors follow formalized processes for reporting on climate finance, such as through the UNFCCC biennial reporting, statistical systems and standards to report mainstreaming of climate finance such as through the OECD DAC Creditor Reporting System, or using dedicated methodologies developed by the MDBs [Multilateral Development Banks] and IDFC [International Development Finance Club]."

593. In 2024, the SCF will prepare the sixth Assessment, which will provide an update on ongoing technical work related to transparency of climate finance, including operational definitions. Specifically, it will include an update to the compilation of definitions of climate finance and criteria used by Parties and institutions, and a comparison of approaches used in the context of reporting climate finance, such as sector-based methodologies, methodologies for estimating mobilized private finance, and domestic climate finance tracking systems.

Figure 8.2.
Data providers, aggregators and reporters of climate finance

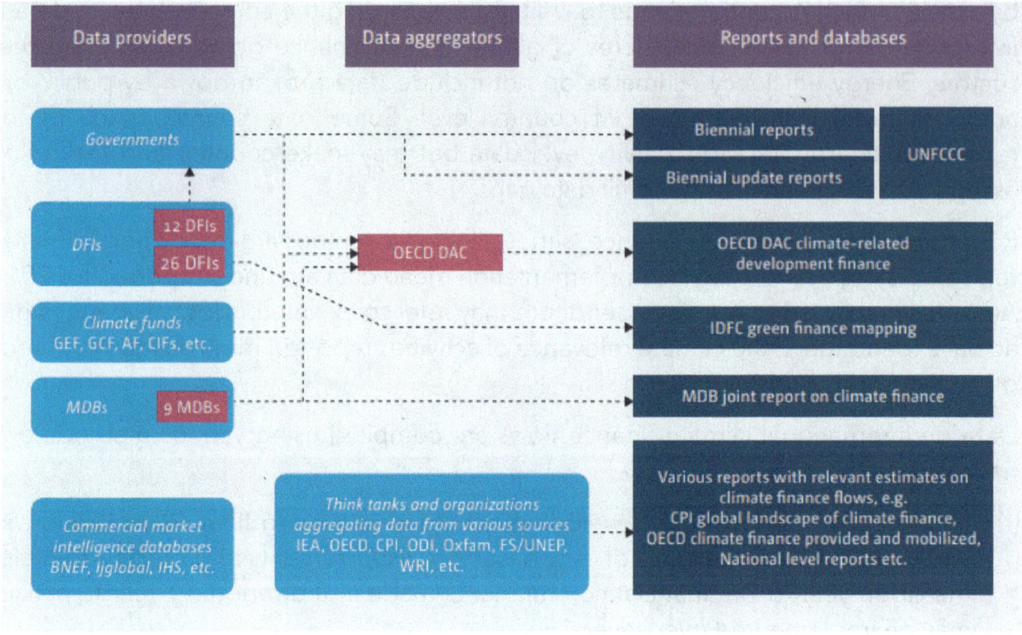

Notes: Dashed arrows indicate formal reporting processes, for example through the UNFCCC, OECD DAC or joint reporting by MDBs and IDFC. Some DFIs report data to their national governments to be included in reporting to the UNFCCC or OECD DAC.

Source: UNFCCC Standing Committee on Finance (2022).

8.2.2 Differentiating between climate, green or sustainable finance

594. Another conceptual issue of interest is differentiating between climate, green and sustainable finance. In 2016, UNEP has delineated the terms of climate finance, green finance, sustainable finance and other related terms in a contribution to the G20 Green Finance Study Group, which aimed to "identify institutional and market barriers to green finance, and based on country experiences, develop options on how to enhance the ability of the financial system to mobilize private capital for green investment". Figure 8.3 shows how these terms relate to environmental and other aspects of Sustainable Development. 'Green finance' is generally used to convey something broader than climate finance, in that it addresses other environmental objectives and risks. It also tends to be understood with a greater focus on greening broad flows of private investment rather than mainly concerning public and public-leveraged financial flows.

Figure 8.3.
Delineation between climate finance, green finance, sustainable finance and other related terms

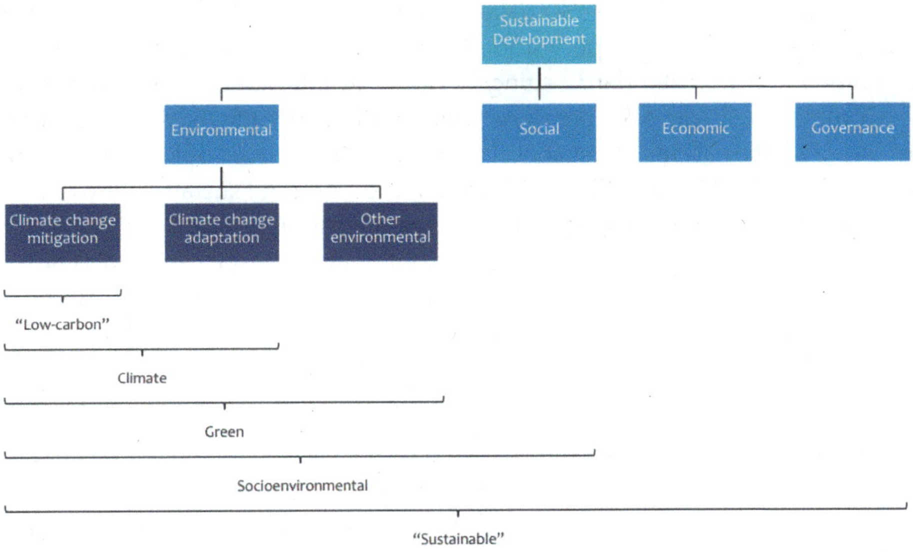

Source: UNEP (2016).

It is also worth mentioning that many renewable energy sources and electrified technologies are now economically on par with the fossil fuel-based equivalents. As the balance tips in favour of the climate friendly options, and supporting infrastructure improves, investment in these technologies will increase without specific climate-related funding (as it will be done on economic grounds rather the climate change grounds). This will represent a significant amount of investment not covered by the current definition of low-carbon finance. There will however be other climate related expenditures (especially adaptation related) that will still require specific funding. Having a mapping between different types of climate-related funding (including the extent to which taxes are hypothecated, and therefore creating a relationship between tax take and transfer/subsidies/etc.) and climate-related expenditures could help tease out these insights and indicate where further development of classifications and data sources would be useful, and possibly help in further defining and delineating between different funding approaches. In compiling statistics that show these relationships, the integration of data will be extremely important.

8.2.3 Broader information needs of the financial sector

595. Climate change poses significant risks to the economy at large and specifically to the stability of the financial sector. At the same time, the financial sector will have to play a role in supporting the transition to a net-zero economy. This includes managing climate-related risks to fiscal and monetary policy and to the financial system thus supporting the green transition. From the perspective of financial stability, it is key to understand the risks resulting from climate change. These include both physical risks and transitional risks. As an example, the banks tend to grant loans to the headquarters of a company. They do not always know the exact location of the establishments/production facilities where these loans are then invested.

596. Financial institutions are increasingly focused on two key aspects of climate change. First, they are examining how it affects the risk and return on their assets. Second, they are considering the influence of climate change on the evolving demand for financial services, catering to both private and public sector clients. In response, financial institutions are actively developing innovative tools and products. They are also accumulating data on transition and physical risks associated with climate change, integrating these insights into their financial decision-making processes. This necessitates adjustments in their business models to stay aligned with these emerging trends. Effectively addressing the dynamic and complex information needs related to the financial dimensions of climate change calls for a multidisciplinary approach, blending expertise from various fields.

597. Understanding the risks arising from climate change is essential for maintaining financial stability. Insurance companies, with their extensive data and experience, are familiar with physical and transition risks, but they are relatively new challenges for other financial institutions, including banks (BIS, Basel Committee on Banking Supervision, 2021). The need to fill firm- and asset-specific data gaps has been highlighted by standard-setting bodies and international organizations, such as the Irving Fisher Committee 2021 and the Financial Stability Board (FSB) 2023. This led to the establishment of the International Sustainability Standards Board (ISSB)'s International Financial Reporting Standards, IFRS S1[126] and IFRS S2[127], focusing on general sustainability-related disclosures and climate-related disclosures, respectively. Besides closing a range of data gaps, financial institutions and supervisory agencies are intensifying efforts to measure vulnerabilities. This includes the development of monitoring metrics and climate scenario analysis (FSB, 2023). The evolving nature of climate risk is reflected in the advancement of regulatory and supervisory practices, including the implementation of climate-specific stress tests and the evaluation of financial institutions' transition plans.

598. In parallel, the importance of climate risk and sustainability considerations is increasingly influencing the business operations of financial institutions through the evolving demands of their customers seeking financial services that align with their sustainability values. This shift in consumer preference is driving banks and other financial institutions to offer green loans, sustainable investment options, and other eco-friendly financial products. Additionally, corporate clients are also adapting to a low-carbon economy, requiring financing for sustainable projects and support in managing climate-related risks in their operations. As a result, financial institutions are not only reassessing their investment portfolios for climate risks but also innovating to meet this new market demand. They are integrating Environmental, Social, and Governance (ESG) criteria into their lending and investment practices, which is reshaping their risk assessment models and business strategies. Furthermore, the growing regulatory emphasis on climate disclosures and sustainable practices is reinforcing this trend, making sustainability a key factor in the competitive landscape of the financial sector. Thus, the demand from customers for sustainability-focused services is compelling financial institutions to embed climate risk and sustainability considerations deeply into their core business models.

599. In addition to their primary role in ensuring financial stability, central banks face climate change impacts in several other key areas: asset and reserve management, financial inclusion initiatives, and the conduct of monetary policy, including collateral policies. The connection between climate change and asset management is similar to the situation for private investors. However, the impact of climate risk on monetary policy is more complex, largely mediated through economic stability. Climate risks, both physical and transitional, can significantly influence economic stability by affecting inflation rates, asset values, and overall economic growth. Physical risks, such as extreme weather events and long-term shifts in climate patterns, can cause infrastructure damage, supply chain disruptions, and commodity price fluctuations. These factors can lead to increased inflation volatility and economic output variations. On the other hand, transitional risks, which arise from the global shift towards a low-carbon economy, could dramatically alter asset valuations, potentially resulting in stranded assets in sectors heavily reliant on fossil fuels. Such shifts necessitate a reassessment of collateral values and credit risks by central banks. Consequently, central banks are now incorporating climate risk evaluations into their monetary policy frameworks, recognizing that climate change has significant internal and systemic implications, and can no longer be regarded as merely an external concern.

600. Climate risk may also significantly impact fiscal policy, as governments worldwide grapple with the economic repercussions of both physical and transitional climate risks. Physical risks necessitate increased government spending on disaster relief, infrastructure repair, and adaptation measures. These expenditures can strain public budgets and divert resources from other critical areas. Transitional risks, associated with the global shift towards a low-carbon economy, also pose fiscal

[126] https://www.ifrs.org/issued-standards/ifrs-sustainability-standards-navigator/ifrs-s1-general-requirements/

[127] https://www.ifrs.org/issued-standards/ifrs-sustainability-standards-navigator/ifrs-s2-climate-related-disclosures/

challenges. Facilitating the transition to a low-carbon economy requires substantial public funding and may lead to shifts in taxation policies, such as carbon taxes or incentives for sustainable practices. Additionally, climate change can impact tax revenues, as sectors vulnerable to climate risks might experience reduced productivity or value, thereby affecting their tax contributions. Governments also face the challenge of supporting communities and industries that are disproportionately affected by climate-related changes, which may necessitate social safety net expansions or targeted economic support. Overall, climate risk compels governments to rethink and restructure their fiscal policies, balancing immediate climate action and adaptation needs with long-term economic sustainability and growth objectives.

8.2.3.1 Network of Central Banks and Supervisors for Greening the Financial System

601. The Network of Central Banks and Supervisors for Greening the Financial System (NGFS) identified climate-related data and data gaps required for various use-cases by financial sector stakeholders (NFGS, 2022). To take stock of available data for financial-sector use cases, the NGFS built a directory providing the types of metrics that are required for each use case. Each stakeholder category and use case is mapped to relevant types of metrics and available data.

Figure 8.4.
Interconnectedness of stakeholders, use cases, metrics and raw data item types

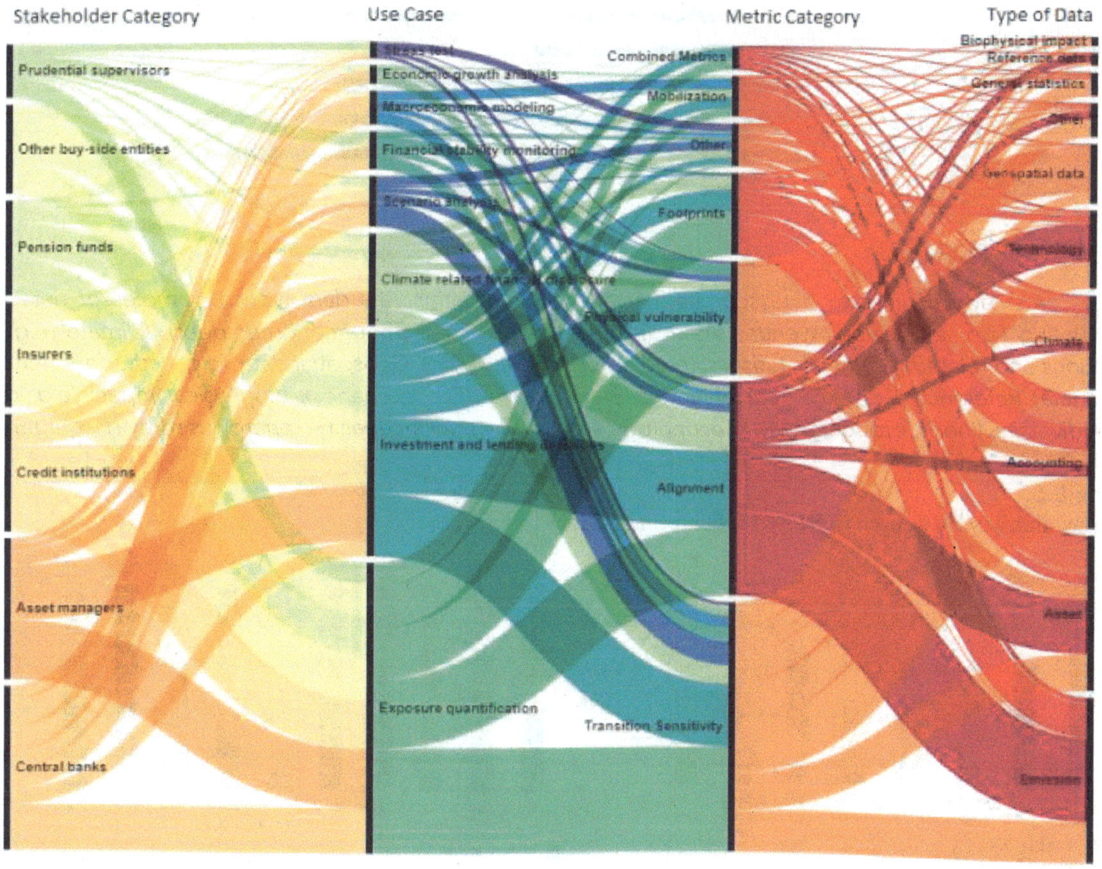

Source: **NFGS (2022).**

602. The directory identified:

(a) Seven stakeholder categories: prudential supervisors, other buy-side entities, pension funds, insurers, credit institutions, asset managers, central banks;

(b) Eight use cases: Exposure quantification; Investment and lending decisions; Macroeconomic modelling; Economic growth analysis; Financial stability monitoring; Climate-related disclosures; Scenario analysis and stress testing

(c) Six types of metrics: footprints, transition sensitivity, physical vulnerability, alignment, mobilization (i.e., scaling up green finance) and combined metrics, plus "Other" - for each metric the data items that are required to compile the metric are identified and links to raw data are provided whenever the data is available

(d) 329 metric/methodology combinations, 1262 raw data items and 748 links to data sources.

603. Through this mapping, the NGFS used the directory to identify data gaps, which are presented on Figure 8.5 as the percentage of missing data across raw data types. Ignoring raw information on accounting and data assets, which often cannot be made available on the directory due to confidentiality reasons, the largest data gaps are observed for biophysical impact, emissions and geospatial data types. For general statistics, data gaps are at 14%.

Figure 8.5
Raw data items without a link to a source/provider by raw data item

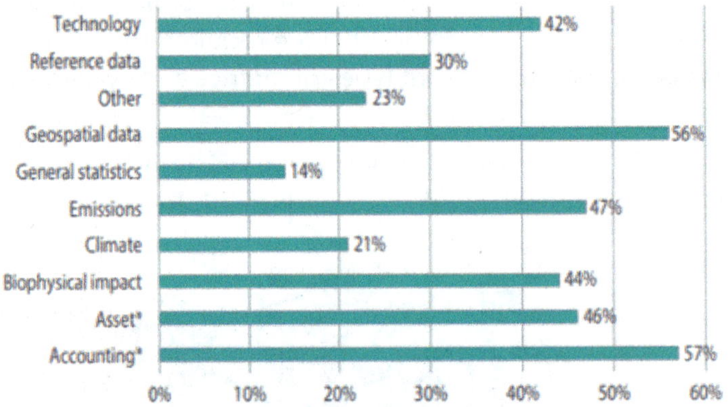

Source: NFGS (2022).

Note: *Percentage of links to data sources that are missing within raw data types. Examples of data types: 'Accounting' (e.g., total sales, revenue), 'Asset' (e.g., asset-level, total amount of green debt), 'Climate' (e.g., CO2 emissions and related), 'General statistics' (e.g., employment, GDP), 'Geospatial data' (e.g., latitude/longitude of asset data), Reference data' (e.g., data relating to IEA Pathways / IPCC scenarios), 'Biophysical impact' (e.g., related to biodiversity, floods, droughts, etc.), 'Technology' (e.g., hydropower capacity', 'energy used by IT equipment')*

Figure 8.6.
Data availability by source (left) and jurisdiction (right).

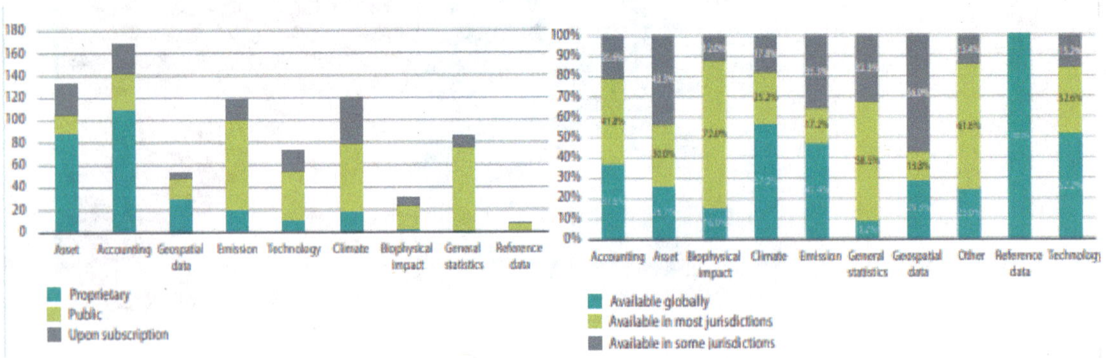

Source: NFGS (2022).

604. Figure 8.6 presents a summary of data availability by source and by jurisdiction. In total, about 1/3 of data items identified in the directory are considered unavailable (29% is either unavailable, unknown or under construction; 28% have accessibility issues). Only less than a quarter of all items in the directory are either official statistics or verified data, while 39% are based on estimations. Among the official statistics items, only less than 10% are available globally, while 32 % are available in some jurisdictions and 58.5% in most.

605. The following challenges in closing the gaps have been identified:

(a) Auditability - needed to build trust and enhance data quality;

(b) Reliance on estimations and modelling - methods that require detailed or specific technical expertise and/or are the proprietary knowledge of private vendors;

(c) Granularity issues - geographical data at entity and asset levels lack specific location information and, where granular data are available, it is often not easily accessible due to cost;

(d) A need for forward looking data for transition risk metrics and physical risk metrics and unavailability of biophysical impact and geospatial data types for the latter;

(e) Capacity building required to understand science-based metrics;

(f) Incompleteness of the information on climate-related data in the directory and the need for further work.

606. The NGFS identified four policy recommendations aimed at producing the building blocks required to bridge data gaps with available, reliable and comparable data:

(a) Foster convergence towards a common and consistent set of global disclosure standards;

(b) Increase the efforts towards mutually shared and operationalized principles for taxonomies and sustainable finance classifications;

(c) Developing well-defined and decision-useful metrics, and methodological standards;

(d) Better leverage of available data sources approaches and tools.

8.2.3.2 Task Force on statistics on sustainable finance and climate related risks

607. Against the background of rising sustainability-related risks for the economy, financial markets and financial institutions, the Committee on Monetary, Financial and Balance of Payments Statistics (CMFB) identified statistics on sustainable finance and climate-related risks as a field of common interest for the European System of Central Banks (ESCB) and the European Statistical System (ESS). To address the topic, the CMFB established in 2020 a task force on statistics on sustainable finance and climate related risks (TF SuFiR), which developed its work in two stages (CMFB, 2023). The Task Force included mainly representatives of Central banks and ministries and the NSOs of two countries.

608. The TF SuFiR followed three objectives: (i) enhancing the utility of existing European data sources by exploring the idea of a "data catalogue", while taking into account other relevant initiatives; (ii) reviewing well established reporting/disclosure registries that could serve as sources for the work on sustainability-related official statistics and for the development of sustainable finance indicators: the EU ETS, the European Pollutant Release and Transfer Register (E-PRTR) and Energy Performance Certificates (as well as related information gathered at national level to monitor implementation of relevant policies); (iii) underlining the importance of better integrating sustainability issues into the classification of economic activities and goods.

609. Regarding the first objective, the TF SuFiR found that the structure of the directory of the Network of Central Banks and Supervisors for Greening the Financial System (the NGFS Directory) can also be used to cover European official data sources, including national official statistics. Consequently, the TF SuFiR offered its findings on the three datasets to the NGFS Directory, which is open to integrating more information on datasets developed by statistical offices and central banks.

610. Regarding the second objective, the TF SuFiR managed, in some cases, to provide inputs to enhance the usability of the aforementioned data sources which were implemented by the European Commission. This underlines the importance of having up-to-date information on legal initiatives for developing data sources. As a follow-up, the TF SuFiR recommended organizing a regular information exchange on legal initiatives in the European Commission so that statistical interests can be articulated and taken into account.

611. Lastly, regarding the third objective, the TF SuFiR recommended that some statistical classifications need adjustment to meet the needs for financial and non-financial statistics on climate-related risks and sustainable finance. The update of the Classification of Products by Activity (CPA) - the EU equivalent of the Central Product Classification (CPC) - is viewed as a timely opportunity to make an important classification fit for this purpose.

8.2.3.3 The third phase of the G20 Data Gaps Initiative (DGI-3)

612. A very important initiative, which also addresses some of the gaps identified by NGFS, is the third phase of the Data Gaps Initiative (DGI-3) by the G20 Finance Ministers and Central Bank Governors.[128] The main objective of the DGI-3 is to address the critical data gaps that exist in the face of the climate crisis, increasing economic polarization, and large-scale digital transformation. Its 14 recommendations are clustered around four statistical areas (i) climate change; (ii) household distributional information; (iii) Fintech and financial inclusion; and (iv) access to private sources of data and administrative data, and data sharing. For each recommendation, policy drivers and target statistical products are identified.

613. Among the 14 recommendations, half are related to climate change:

Table 8.1.
Climate-related recommendations in the third phase of the G20 Data Gaps Initiative

Recommendation	Target statistical output
1: GHG Emission Accounts and National Carbon Footprints	Annual Air Emissions Accounts (for GHGs) by industry and estimates of National Carbon Footprints by demand category (exports, household consumption etc.)
2: Energy Accounts	Annual Energy Accounts
3: Carbon Footprints of Foreign Direct Investment (FDI)	CO_2 emissions per unit of output of foreign controlled multinational enterprises and domestic controlled enterprises by industry
4: Green Finance (Green Debt and Equity Securities Financing)	Issuances and holdings of green debt securities and listed shares securities
5: Forward Looking Physical and Transition Risk Indicators	Forward-looking physical and transition risk indicators related to populations, assets, output/production, firm profitability, and wealth
6: Climate-Impacting Government Subsidies	Annual estimates of government subsidies by type of subsidy
7: Climate Change Mitigation and Adaptation Expenditures	Annual estimates of current and capital expenditures on climate mitigation and adaptation activities by sector (Governments, nonprofit institutions serving households, households, nonfinancial corporations, financial corporations)

Source: IMF.

614. DGI-3 Recommendation 4 on Green Finance and Recommendation 5 on Forward-Looking Physical and Transition Risk Indicators are particularly relevant for the financial sector. The financial sector plays an important role in responding to the financing needs associated with climate change mitigation, adaptation, and transition.

615. The aim of the DGI-3 Recommendation 4 is to address data gaps by developing reporting templates and the associated methodological guidance to collect experimental data on issuances and holdings of green debt securities and listed shares securities consistent to the extent possible with the Handbook on Securities Statistics.

[128] Documents related to the DGI-3 can be found at https://www.imf.org/en/News/Seminars/Conferences/DGI/g20-dgi-progress-reports-and-other-documents

616. The DGI-3 Recommendation 5 focuses on forward-looking climate related physical and transition risks. The forward-looking assessment of climate-related physical risks is essential, considering the rising frequency and severity of natural hazards such as floods, droughts, fires and extreme heat. Additionally, assessing how climate policies will affect populations, national wealth, and firms while transitioning to a low-carbon future is crucial for anticipating and addressing the impacts of climate change on the economy, and in turn, on the financial sector. Despite the increasing awareness of these risks, a significant challenge is the lack of comprehensive data and a structured framework for measurement, preventing financial institutions to comprehensively assess climate related risks. Thus, the DGI-3 aims to close these critical policy-relevant data gaps with the support of G20 economies.

8.3 How NSOs can contribute

617. Statistics related to the financial aspects of climate change is an emerging area of work for official statisticians. At the 11th Expert Forum for Producers and Users of Climate Change-related Statistics[129], held in 2023, the following elements of statistics related to financial aspects of climate change were mentioned:

(a) GHG emissions from companies;

(b) Footprint of companies;

(c) Consumption of fossil fuels by companies;

(d) Exposure of companies to extreme events;

(e) Subsidies/transfers given by government;

(f) Subsidies received by companies/households;

(g) Public funds related to climate change;

(h) Expenditure of government/companies to mitigate climate change;

(i) Expenditure of government/companies to adapt to climate change;

(j) Support of the government to developing countries;

(k) Impacts (climate change and larger social) of project finance by climate funds.

618. There are many stakeholders in this topic including:

(a) NSOs;

(b) Central banks;

(c) Financial sector supervisory authorities;

(d) Ministries of finance and treasury;

(e) Energy and environment ministries;

(f) Financial sector companies (banks, insurance companies, etc.);

(g) Stock exchanges.

619. The topic is large, and different actors can lead on specific issues, depending on the existing division of responsibilities in each country. For example, data on international flows may often be in the hand of development ministries or development banks, data on finance in the private sector could be developed by national central banks, and data on public sector are often under the responsibility of ministries of finance. But there are also some issues for which typically NSOs are in a good position

[129] https://unece.org/statistics/events/EFCCRS2023

to lead. Each issue should be allocated to a primary agency in each country and the NSO is well-placed to start that conversation and coordinate the work.

620. Depending on the above, NSOs can be involved in some of the following:

(a) Supporting the coordination/cooperation;

(b) Monitoring climate finance;

(c) Producing statistics on government expenditures, subsidies, transfers and investments;

(d) Measuring climate impacts of economic/financial activities;

(e) Producing data for assessing physical and transition risks;

(f) Assisting with the efforts towards harmonization, through implementation of revised statistical standards and classifications, allowing for capturing climate change-related activities;

(g) Promoting statistical indicators by including them in a climate change related indicator dashboard produced by the NSO;

(h) Implementing some of the DGI-3 Recommendations.

8.3.1 Coordination and cooperation

621. Taking into account the need of a close cooperation among different Institutions such as central banks, ministries, NSOs, environmental agencies and private stakeholders, establishing a committee/working group could be helpful to decide on the way forward, establish priorities and agree on work calendars. Setting up such a committee should also help establish the links and improve coherence between data used for international reporting under the Paris Agreement, data used by the financial sector and data published for the public.

Netherlands - Platform for Sustainable Finance[130]

In the Netherlands, the Platform for Sustainable Finance was created to promote and encourage awareness of sustainable finance in the financial sector. It is a partnership between the Dutch financial sector, government and supervisory authorities in which they work together on promoting sustainable finance in the Netherlands.

The Platform members periodically select priority themes based on signals from the working group and presentations of new themes by discussants. The Platform focuses on two types of themes. The first type are new sustainability themes that are still in an early stage of development in the financial sector or government. For these themes, the main objective is knowledge acquisition and exchange- The second type of themes are already recognized issues, or that will inevitably become issues, along with the question of how the sector can best respond to them. For these types of themes, the objective is to encourage action and formulate recommendations for the various actors.

8.3.2 Identifying and filling data gaps

622. NSOs also have a role in identifying and filling data gaps, building on the work of existing international groups discussed in the chapter. These groups have specific needs in mind, each requiring input data to be collected. If not effectively coordinated, this is might result in duplicated collection of data and leaving unfilled the data gaps impacting on lower-resourced needs. NSOs have expertise in collating multiple user needs and designing data collection that can meet those needs efficiently. NSOs also have expertise in procuring data (as an independent trustworthy institution) and in designing flexible and large data systems and statistical infrastructure. NSOs should lead the process of designing the data system to meet multiple needs from minimum data sources. Where there is scope to design (or redesign) data collections NSOs can ensure the data is collected correctly to meet the widest range of needs possible.

[130] See https://www.dnb.nl/groene-economie/platform-voor-duurzame-financiering/ and https://www.dnb.nl/media/cegfouuq/vision-strategy-pvdf.pdf

8.3.3 Monitoring climate finance

623. As explained in chapter 3, developed country Parties will need to report on financial support provided and mobilized under the ETF. However, with the growing volumes of climate finance, in addition to the official reporting, it becomes increasingly important to publish trusted statistics about climate finance and its impacts for the general public.

United Kingdom - Statistics on the impacts of international climate finance[131]

International Climate Finance (ICF) is the UK Government's primary instrument to deliver the UK's share of the financial and technical assistance to developing countries under the UNFCCC $100bn per year climate finance commitment. UK government has developed a set of Key Performance Indicators (KPIs) to support monitoring of positive impacts, or results, that its finance is delivering in ODA-eligible countries. Each year, the Foreign Commonwealth and Development Office of the United Kingdom publishes a set of statistics which show all the results that have been achieved in the preceding year, as well as a time series back to 2011 when climate finance started to be monitored in this way. The main users of these statistics are the general public, in particular providing transparency about what government funds are being spent on. The climate finance results are not designated as official statistics, but, where possible, the involved departments voluntarily apply the UK Statistics Authority's Code of Practice for Statistics in their production. Data sources include routine monitoring by implementing partners, management information systems of partner country governments and household surveys. For the purpose of this analysis, the UK international climate finance are defined as OECD DAC Official Development Assistance and mobilized private finance, with climate change adaptation and mitigation Rio Markers (principal or significant) (see Figure 8.7). Table 8.2 presents the UK international climate finance results achieved from April 2011 to March 2022.

Figure 8.7.
United Kingdom international climate finance defined as OECD DAC Official Development Assistance and mobilized private finance, with climate change adaptation and mitigation Rio Markers (principal or significant)

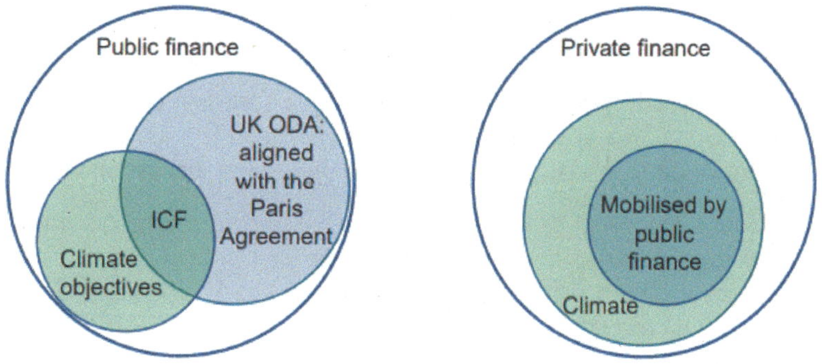

Source: ONS

Table 8.2
UK International Climate Finance results achieved from April 2011 to March 2024

ICF KPI	KPI Name	Total achieved
KPI 1	Number of people supported to better adapt to the effects of climate change	110,075,000 people
KPI 2.1	Number of people with improved access to clean energy	82,132,000 people
KPI 2.2	Number of social institutions with improved access to clean energy	1,111 institutions
KPI 4	Number of people whose resilience has been improved	32,521,000 people
KPI 6	Net Change in GHG Emissions reduced or avoided	105,720,000 tCO_2e
KPI 7	Installed capacity of clean energy	4,561 MW
KPI 8	Area where deforestation has been avoided	759,000 hectares
KPI 10	Value of ecosystem services generated or protected	£5,630,000

[131] https://unece.org/statistics/documents/2023/08/presentations/statistics-impacts-uk-international-climate-finance and https://www.gov.uk/government/publications/uk-climate-finance-results-2022/uk-international-climate-finance-results-2022

KPI 11	Public finance mobilized for climate change purposes	£8,412,447,000
KPI 12	Private finance mobilized for climate change purposes	£7,853,652,000
KPI 15	Extent to which ICF intervention is likely to lead to transformational change	62.5% programmes scored a 4 or 5
KPI 17	Area of land that has received sustainable land management practices	4,229,000 hectares

***Source*: UK international climate finance results 2024**[132]

8.3.4 Statistics on government expenditures, subsidies and transfers

624. Reporting on the distribution of funds for climate change mitigation and adaptation is usually the responsibility of environment or finance ministries. However, much of this information is also available in the accounts of environmental subsidies and transfers compiled by NSOs, provided that the compilation of these accounts through environmental classification is sufficiently detailed.

625. These accounts have the advantage of disaggregating subsidies and transfers by donor as well as by recipient. Compiling these accounts is a complicated operation requiring a great deal of information. In view of the importance of these accounts in assessing the cost of the transition to a low-carbon and resilient society, NSOs should make the necessary resources available for their compilation.

626. Another interesting account developed by NSOs are the environmental tax accounts, which present a picture of government efforts in using this financial instrument to influence consumer choice.

627. Governments, companies and households are making expenditures that can be seen as concrete actions to mitigate their emissions and adapt to climate change. Following the example of the culture and sports sectors, it is possible to characterize a climate change mitigation and adaptation sector. This sector is partially contained within the environmental goods and services sector.

Ireland - statistical release on climate-related subsidies

As part of the implementation of SEEA, Eurostat has developed a series of environmental accounts modules, including the Environmental Subsidies and Similar Transfers module. The Central Statistics Office in Ireland submits data for this module annually and publishes an annual national statistical release.

CSO approach to the module is to collect data on environmental transfers from government departments and agencies using a bottom-up approach. As per Eurostat guidelines, each subsidy or grant scheme is then classified according to type of transaction (European System of Accounts (ESA 2010) code), environmental domain (Classification of Environmental Protection Activities and Expenditure (CEPA) / Classification of Resource Management Activities (CReMA))[133], and NACE and institutional sector of the beneficiary.

In addition, for the statistical release, CSO tags each climate-related scheme and estimates the total annual amount paid in climate-related subsidies through the use of Rio Markers. This means that if the principal objective of the programme was climate protection, then the full amount of the subsidy is included (100%), while if one of a number of significant objectives was climate protection then it is partially included (40%).

Some schemes which fall under the energy efficiency, renewable energy or forestry CEPA/CreMA domains also have climate protection as a major objective, resulting in 40% of payments under these programmes being included as climate-related through the method outlined above. CSO includes a list of climate-related subsidy schemes in the background notes to the release, along with the Rio Marker applied to each scheme. For a small number of these schemes, the full amount (100%) of the subsidy is included: some examples are climate research funding, the peatlands climate action scheme and the Irish government's contribution to the UN Framework Convention on Climate Change.

There are plans to review this methodology for next year's release in light of the relevance of certain other schemes such as retrofitting grant programmes to the Climate Action Plan. For example, under the current

[132] https://www.gov.uk/government/publications/uk-international-climate-finance-results-2024/uk-international-climate-finance-results-2024

[133] The CEPA and CReMA classifications are now being replaced by the new Classification of Environmental Purposes - see paragraph 634(c).

approach 40% of funding provided under the new National Home Retrofit Scheme would be counted as climate-related expenditure. The CSO methodological review will look at whether the full amount paid under this scheme should be included.

8.3.5 Measuring climate impacts of economic/financial activities

628. Depending on the country, energy consumption and GHG emissions data are usually collected by the NSO, or by the Ministry of Energy or the Ministry of the Environment (or their respective administrations). In the absence of observed data, emissions are estimated from energy consumption by branch of activity as part of energy balances and physical accounts of energy flows. Estimates of GHG emissions, on the other hand, are produced as part of the compilation of air pollutant accounts. If the NSO is a custodian of the centralization and archiving of micro-data, they can collect and store these statistics.

629. As described in Chapter 3, energy and GHG emissions data are important data sources to compile GHG inventories. They are also extremely useful for financial institutions in calculating financed emissions. They can also be used for consumption-based indicators such as carbon footprints.

630. The recent ESG directive requires certain companies to provide additional reporting on environmental data such as energy consumption and GHG emissions. These extra-financial reports are a new source of climate change-related statistics for which NSOs could play a centralizing role. The new data will not only improve national statistics but would also be useful to the financial sector in characterizing financed emissions.

8.3.6 Producing data for assessing physical and transition risks

631. The financial sector and political leaders are particularly interested in obtaining an assessment of the exposures and vulnerability of the population and businesses to the risks associated with climate change. To this end, some NSOs are working to maintain databases on extreme events and disasters, on the populations affected (or potentially affected), broken down by various socio-economic characteristics (age, sex, etc.), and on companies located in high-risk areas. This information is generally reused in national emergency plans but could also be used by the financial sector to assess local risks to companies financed by the sector.

8.3.7 Contributing to updates and implementing revised statistical standards

632. The UN Statistical Commission is expected to endorse the System of National Accounts 2025 (2025 SNA) in the first quarter of 2025. One of the priority areas for the update of the 2008 SNA is Wellbeing and Sustainability, which includes many topics relevant for climate change statistics, including statistics on climate finance. The proposed changes to the SNA address, for example, accounting for sustainable finance, the recording of emission trading schemes, improvements to statistics on financial risks, and better recording of natural capital.[134]

633. In addition to the SNA (and the Balance of Payments Manual, which is updated in parallel), an update of the Government Finance Statistics Manual (GFSM) and the SEEA Central Framework is expected to start in the coming years.

634. There has also been considerable progress in revising critical statistical classifications, which can facilitate statistics on climate finance:

(a) The United Nations Statistical Commission endorsed the structure of revision 5 of the International Standard Industrial Classification of All Economic Activities (ISIC Rev.5) in 2023. The classification now allows the distinction of important activities related to climate change

---

[134] Report of the Intersecretariat Working Group on National Accounts on the recommendations for the update of the 2008 System of National Accounts, to the UN Statistical Commission in its 55th session, February-March 2024. https://unstats.un.org/UNSDWebsite/statcom/session_55/documents/2024-9-NationalAccounts-Recommendations-E.pdf

and environmental protection and conservation. For example, the inclusion of carbon remediation, capture and storage activities in Division 39 Remediation and other waste management service activities. ISIC now also distinguishes between electric power generation activities from non-renewable and renewable sources.

(b) Version 3 of the CPC now puts greater prominence on environmental services. For example, by creating a new class for carbon capture and storage services. The United Nations Statistical Commission endorsed the new structure of the CPC in 2024.

(c) The Classification of Environmental Purposes (CEP), endorsed by the UN Statistical Commission in 2024, combines the Classification of Environmental Protection Activities and the Classification of Resource Management Activities to organize the statistics produced within the framework of the environmental monetary accounts. The CEP classification clearly includes virtually all activities that are implemented to mitigate climate change. On the other hand, the majority of activities relating to adaptation to climate change (combating drought, flooding, etc.) are excluded. The new CEP therefore only partially meets the NSOs' need for a harmonized classification that can be used to produce statistics covering, in a harmonized way, all subjects relating to climate change.

635. Another highly relevant workstream for statistics on climate finance is the revision of the Classification of the Functions of Government (COFOG), which will impact how government expenditure can be measured. The current version of the classification is inadequate to capture climate change spending, as it is encompassed by broader sub-categories. A Task Team of the Committee of Experts on International Statistical Classification was established in March 2024 to advance this revision.

636. Applying classifications is at the heart of NSOs' work. To extract relevant information for policymakers and the financial sector, it is essential that statistical standards and classifications be adapted to enable capturing activities related to climate change. Contributing to these update processes and implementing the revised standards and classifications are important opportunities for NSOs to contribute to the work on statistics on climate finance.

8.3.8 Data integration

637. Given the role expertise of NSOs in classifications and frameworks, and their development of registers and geospatial infrastructure, NSOs can be expected to play a key role in integrating data for compiling statistics showing the relationships between different types of climate-related funding.

8.4 Conclusions and recommendations

638. The chapter introduced, in a non-exhaustive manner, several issues related to financial aspects of climate change that should be of interest to the official statistics community. These range from defining, tracking and measuring climate finance through differentiating between climate, green, and sustainable finance and describing the information needs of the financial sector.

639. Many initiatives are already in place in this thematic area, from formal bodies under the UNFCCC, other international initiatives such as DGI-3 and task forces in the central bank/financial sector community. At the same time, many gaps and challenges persist and many recommendations about the need for further work have been made.

640. Potential areas where NSOs can contribute include coordination/cooperation, statistics on government expenditures, subsidies and transfers, supporting national climate finance reporting, measuring climate impacts of economic and financial activities, producing data for assessing physical and transition risks, supporting efforts towards standardization, and measuring climate investment. A contribution of statistical offices is clearly needed to solve some of them, while for others, NSOs may mainly be keeping a strategic and coordinating perspective. SEEA framework can play an important role in bringing the relevant information into a common and consistent framework. The ongoing

revisions of the SEEA-CF, GFS and COFOG provide the opportunity to make these frameworks even more relevant for climate change. Responsibilities for these statistics vary from country to country and the role of the NSO can vary accordingly and different actors can lead on specific issues. Regardless of the institutional arrangements in place, it should be a collaborative effort.

641. Many topics discussed in this chapter require further international work, which should build on existing initiatives and collaborations. More discussion on how to move forward is needed. A decision on what to prioritize should take into account the potential speed of work and involve experts from economic statistics and national accounts.

642. At this stage two concrete recommendations can be made:

(a) A national coordination committee, including the NSO, should be established to facilitate cooperation, avoid any overlaps to best leverage the work that NSOs already do;

(b) A key area that should be exploited in the very short term is facilitating data access primarily for statistical compilation purposes, e.g. by the central banks, but as well also for research. This might include locational data and business registers.

Chapter 9 Guidance on cross-cutting issues

9.1 Governance and coordination

643. As presented in earlier chapters of this report, there are many institutions producing statistics and data relevant for climate change policymaking, both within and outside of national statistical systems. Within the NSOs themselves, the production of statistics and indicators mentioned in previous chapters may require cooperation among various domain experts and leveraging of innovative methods and new data sources.

644. Chapter 3 explained the importance of establishing sustainable institutional arrangements for climate reporting. The same requirement can be extended to producing climate change-related statistics in general, both for policymaking and informing the public. Especially when data needs are many and urgent, data sources distributed and resources scarce, it is important to establish clear mechanisms for collaboration and coordination to avoid duplications of efforts, capitalize on synergies and enable cross-domain learning.

645. The specificity of this domain, with many relevant institutions with established mandates and multiple data sources, makes it particularly suitable for being approached from a data stewardship perspective, meaning "*ensuring the ethical and responsible creation, collection, management, use and reuse of data so that they are used for public good and benefit the full community of data users*" (UNECE, 2023a). Through their expertise in data management, processing and safeguarding, NSOs can play a vital role in ensuring that administrative data are treated as data assets and their value for informing climate policymaking is captured (see also examples of use of utility microdata and climate data rescue in Ireland, in sections 4.3.2 and 5.3.2.6 respectively).

Ireland – data stewardship in environmental domain[135]

Ireland has a highly centralized statistical system, with the Central Statistics Office (CSO) responsible for 82% of European statistics. The environment domain is different, since statistics required to meet many European legislative requirements are produced by the scientific agencies who have regulatory responsibility. Of the 14 'Other National Authorities' reporting European statistics, 3 are environmental and 3 more are in related fields such as sustainable agriculture, food production and fisheries. In addition, while many Public Sector Bodies (PSBs) produce statistics as a by-product of their administrative functions, statistical production methodologies are much more complex for environmental statistics, such as the European Energy Balance Sheets. These more complex statistics require multi-agency collaboration, including CSO involvement. Data sharing between public agencies is quite limited in Ireland and is usually limited to defined operational purposes rather than, e.g., for multidimensional policy analysis. Finally, much of the data required for environmental statistics is held within the public sector in Ireland, such as climate sensor and utility meter data.

In addition to facilitating data sharing, there also needs to be a National 'Data Steward' who co-ordinates data skills, services and standards in the public sector so that all Public Sector Bodies can benefit from enhancements in the 'data ecosystem'. The CSO is willing to take on this role, in collaboration with others, and would offer a 'catalogue of services' for PSBs varies depending on the context in each policy domain. In the environment domain, the CSO can potentially provide the services described below:

Partnerships. The CSO has engaged in a number of strategic partnerships in recent times with agencies that have complementary technical skills. A successful partnership has been established with the national mapping authority in Ireland, Ordnance Survey Ireland (OSI). Partnerships with the two key Irish environmental agencies, the Environmental Protection Agency and the Sustainable Energy Authority of Ireland, are deepening, through the continued work of long-standing liaison groups.

Secondments. CSO professional statistical staff have been seconded to PSBs on request for more than 30 years with the seconding PSB paying for the cost of the service. The seconded statistical staff provide a broad range of statistical services and are also a liaison for services being provided by the CSO within the seconding organization.

[135] See more in Morrin, P. (2022).

Bespoke and regular multidimensional analysis based on data linkage. The CSO has long standing access to identifiable data from across the administrative system which facilitates matching based on identifiers for people and properties where available. Where identifiers are not available, matching based on deterministic methods is possible with high success rates where names, sex and dates of birth, in particular, are recorded accurately. The service CSO provides across all domains, in addition to official statistics, includes 'pathfinder projects' and short-term project resources via the Data Science and Statistical Support Unit.

Researcher access. Recently, there have been demands from the policy side, particularly health and education sectors, for the CSO to link and integrate relevant data sources, and to subsequently provide research access to these newly created Research Microdata Files within the framework of the Statistics Act, 1993. Enhanced governance around these Research Microdata Files implemented in the context of COVID data access for researchers is the basis for approaches in other domains. The service is already provided for a small number of research projects funded by environmental agencies and there is potential for expansion in the context of European Data Spaces for the Green Deal and the rapid expansion of research into environmental issues.

The Netherlands[136]

The European Green Deal (EGD) includes an ambitious set of interwoven action plans in various policy areas. In order to be as responsive as possible to the data needs of key EGD stakeholders at the European, national and regional levels, Statistics Netherlands is setting up a dedicated EGD programme. Under this programme, Statistics Netherlands will set up a national statistical EGD Action Plan, organize data-articulation meetings with policymakers and others, disseminate an EGD Dashboard/portal and intensify cooperation internally across departments and externally with other data producers and other National Statistical Institutes for sharing experience and best practices.

9.2 Engagement with policymakers

646. Many information needs described in this document stern from the priorities defined in national strategies or climate action plans. Development of statistics takes time, so it is important that NSOs are aware of the upcoming data demands early and can take them into account in planning their work.

647. Active engagement with policymakers can also help to bridge the gap between data and insights - policymakers may not easily translate their questions into data requirements - either to be produced or already available, increase the statistical literacy of users and improve the understanding of the policy questions by the statisticians. It is therefore important to ensure that various fora for such exchanges are available.

648. NSOs often have established mechanisms for interacting with policymakers and other user representatives, e.g. through a general statistical advisory council or a process of formulating statistical programmes. These bodies and mechanisms can be very useful for engaging with users, understanding user needs and deciding on priorities in the climate change-related statistics domain, provided that climate policy users are participating in them.

649. On the other hand, there might also be special bodies, groups or committees dedicated to interdepartmental coordination of climate policy responses. In those cases, it is important that NSOs are represented in such groups, even if they do not have a formally assigned role, e.g. in the climate reporting process. Other relevant avenues for interaction may be within groups working on national SDG implementation and reporting, user fora, webinars or informal gatherings.

Azerbaijan

On 11 March 2020, the State Commission on Climate Change was formed by the Decree of the President of Azerbaijan Republic, to intensify activities to fulfil relevant commitments of the Republic Azerbaijan under UNFCCC, as well as the Paris Agreement. The State Statistical Committee is among the members of the Commission and regularly takes part in the meetings organized by the Commission.

[136] See more in Werkhoven, T. (2022).

The Netherlands

Investing in relationships with stakeholders. Statistics Netherlands has appointed account managers for all relevant stakeholders (e.g., Ministries, agencies, and research institutes), who facilitate the process of providing the requested statistical data and expertise to these stakeholders. Their mission is:

(a) Build and sustain relationships;

(b) Focus on visibility and awareness;

(c) Propagating the multiannual program;

(d) Switching from demand-driven to need-driven and from custom-made statistics to co-creation;

(e) Setting up plans, teams and meetings for the account with a view to a long term strategy.

Figure 9.1.
Internal and external role of stakeholder relationship management in Statistics Netherlands

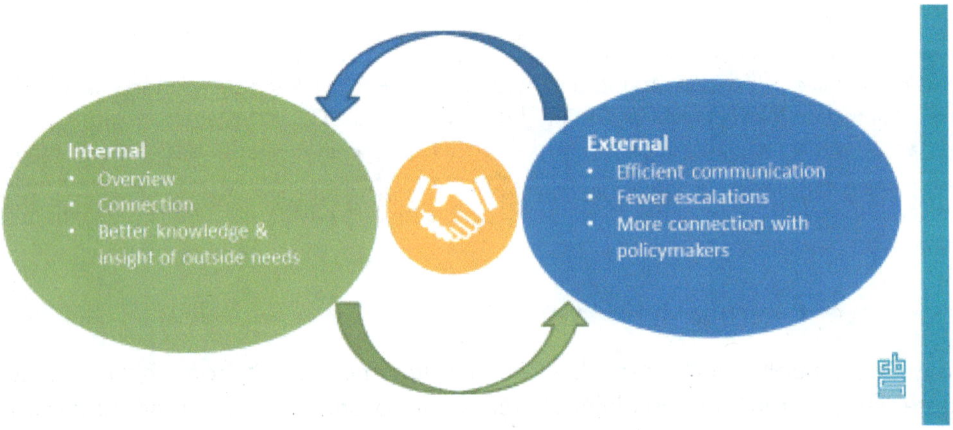

Source: **Statistics Netherlands.**

In addition to account managers, line managers at various levels (heads, directors, the Director General themself) also maintain relationships with data sources and data users, including researchers. They all see the value and play their part in building up a network and promoting visibility and awareness of the statistical treasury. Everybody is an ambassador of the organization and of the data.

Statistics Netherlands is seeing positive results from the investment in relationships: internally and externally, for the short term and the long term. For example, the account managers gain a unique view on the organization and society and awareness of external developments. Also, the employees involved must balance building up close relationships with compliance with fundamental statistical principles and the statistical law, as NSOs do not meet needs at any costs, and in some circumstances may need to refuse some requests.

Engagement with local authorities - VIVET www.programmavivet.nl (in Dutch). In 2019, The Netherlands adopted a National Climate Agreement. This was a result of a compromise between many societal partners from government, industry, NGOs, etc. The goals of the agreement implied a need for policy, planning and monitoring on both a national and decentral level. In many cases, this has led to new information needs for not yet available monitoring data, and the question whether to accept or fill these data gaps.

Generally said, there was a need for improving the provision of information to local authorities related to the energy transition. For this, a broad cooperation between central government, local government, special units and data institutes was formed in 2019, called VIVET.

The central idea behind VIVET is that all parties decide on what questions to answer with a focus on usefulness for municipalities. The central government provides funding, while data institutes execute projects to get new data. Furthermore, it is essential that a party is involved in each project that guarantees the financing if it is decided to switch to monitoring after completion of the project.

Examples of projects are:

(a) Improving existing indicators;

(b) Standardisation of reporting on landbased windmills and photovoltaics;

(c) Data on urban heating;

(d) Location of private electric vehicle charging points;

(e) Investigation on data needs for enforcement.

What sets VIVET apart from other initiatives is the very cooperative atmosphere, the demand-driven approach and the cultivation of an eco-system where people and organisations know each other and work together. Twice a year, a network event is organised with 100-150 participants (in a country with about 340 municipalities). Challenges are the varying levels of data-skills of end users and, most importantly, formulating well-articulated questions.

United Kingdom

The ONS engages with policy colleagues and other stakeholders through:

(a) **Policy Liaison Unit:** This team within ONS operates as the link to central government policy areas, seeking to maximize the opportunities for ONS data and analysis. The Unit operates flexibly across policy areas and organizations to understand the needs of policymakers and ensure that decisions are informed by the latest and most relevant evidence and analysis;

(b) **Events**: ONS organizes seminars for certain publications, such as our Natural Capital Accounts, to disseminate findings and engage directly with users. In July 2023, a major event was organized for the relaunched Measuring National Well-being dashboard, as part of the ONS engagement around ongoing "Beyond GDP" work;

(c) Advisory boards and steering groups;

(d) Strong interdepartmental relationships;

(e) Public consultations;

(f) **Government Departments:** ONS engages strongly with government departments on our outputs with user groups.

9.3 Strengthening data collection

650. Reliable statistics and indicators require sound data, and many areas within the climate change statistics domain would benefit from strengthened data collection. Traditional data collection methods, such as censuses and surveys, can be examined from the perspective of meeting climate change-related data needs. Some may already include data that could be used, while in others a possibility of including additional questions or modules could be considered.

651. New data sources, such as administrative data, Earth observations or electric meters, can complement traditional methods, such as surveys, to increase the efficiency of data production or enable the production of new statistics and indicators.

652. Both approaches require, in various proportions, additional financial resources, expertise, IT infrastructure, user demands, justification for new data collection and high-level support.

653. **Administrative microdata.** Among new data sources for climate change, energy and environment statistics, particularly promising are administrative microdata, such as utility meter data, energy performance certificates of buildings and vehicle test data. Access to administrative microdata relies on several frameworks being in place, such as legal, policy, organizational and technical frameworks. A key question for obtaining access may be the status of the entities collecting the proposed microdata, which may vary by country and microdata category, e.g. single public authority, multiple public authorities, local or national entities, private entities, etc. (Central Statistics Office of Ireland and UNECE, 2023).

654. The *UNECE Guidance on Modernizing Statistical Legislation* (2018) recommended that the law on statistics should include provisions that oblige all national and local authorities and private bodies to provide data in their possession to the producers of official statistics if needed for statistical purposes and as far as appropriate within the policy settings of the public sector of the relevant

jurisdiction. Data should be provided free of charge, at the level of detail necessary for the production of official statistics (including identifiers) and with the metadata that enable the assessment of data quality. The Guidance also recognized that there are opportunities for far greater use of administrative data and new data sources to shed light on complex policy problems and to reduce the costs and burden caused by direct data collection. Legislation must give statistical offices the ability to overcome challenges, acquire various data for statistical purposes and allow wide collaboration with public and private partners.

655. The EU Regulation 223/2009 on European statistics is currently under revision to take account of the developments that have transformed the data landscape and the environment in which European statistics are developed, produced and disseminated. Among others, the intention of the revision is to seize opportunities provided by digital transformation - new data sources, emerging technologies and the growing data ecosystem - to meet the growing demand for more up-to-date, granular and responsive official statistics to support EU policies and public debate.

656. In addition to legal frameworks, policy, organizational and technical frameworks are also relevant for obtaining access to data. Policy frameworks that will have an influence on the right of access may include general policies on data sharing within the government, including national or international codes of practice. Organizational arrangements may cover putting in place written agreements, such as contracts, service level agreements or memorandums of understanding. Finally, technical frameworks include technical means for transferring the data.

657. Establishing partnerships and showcasing the policy relevance of the statistics that can be produced from source data can also help in obtaining access.

658. **Existing data collections.** In some cases, strengthening data collection may require introducing some changes to existing data collection activities, such as population and housing censuses or household surveys.[137]

Ireland

The Statistics Act, 1993, permits the CSO to request confidential microdata from public authorities for statistical purposes. Under the Act, the CSO obtained access to electricity, gas and water meter data. The Act was used to gather a lot of other environment-related information such as energy efficiency audits, vehicle odometer readings, Emissions Trading Scheme, Large Industry Energy Network, and Public Sector Energy Efficiency schemes.

The comprehensiveness of administrative microdata gives it the potential to measure progress towards sectoral targets without having to undertake new statistical surveys, so long as the microdata cover a high proportion of household and enterprise populations. Electricity meters are very comprehensive and can be used for many additional specialized purposes, such as analysis of vacant and new dwellings and maintenance of a business register. In Ireland, gas and water meters typically cover urban areas and customers located close to the main pipelines. Based on these microdata, the CSO publishes, e.g.:

(a) An annual statistical release showing consumption by user group and quarter for detailed geographical areas. The file was used to identify data centres by doing searches using customer name, address and consumption level. The CSO published a thematic report which showed that data centres used more electricity in 2021 than all rural households combined; [138]

(b) An annual statistical release showing gas consumption by user group and quarter at a detailed level of geography;[139]

(c) An annual statistical release showing water consumption by month at a detailed level of geography. A statistical analysis by the CSO showed that the highest decile of metered households accounted for 36%

[137] Based on a recommendation emanating from EGES, a sub-group was established at UNSD to develop a core set of climate change questions which can be included in censuses and surveys for countries to collect data for the Global Set where appropriate. At 10th EGES meeting, examples of including such questions in population and housing census were presented by Tanzania (https://unstats.un.org/unsd/envstats/fdes/EGES10/2.3.3%20Tanzania_experience_PHC.pdf) and in household surveys (https://unstats.un.org/unsd/envstats/fdes/EGES10/2.3.2%20SPC%20Developments.pdf) by the Pacific Community.
[138] https://www.cso.ie/en/statistics/energy/datacentresmeteredelectricityconsumption/
[139] https://www.cso.ie/en/statistics/energy/networkedgasconsumption/

of household consumption, indicating that there is significant wastage due to leaks between the meter and the dwelling.[140]

9.4 Need for local and geospatially enabled data

659. Improving availability of granular/local and geospatially enabled statistical data has been identified as an important contribution of NSOs for informing climate change mitigation, adaptation and just transition, and informing the public about climate-related issues.

660. The effectiveness of indicators and of data depends not only on the statistical design of the data, but also on an adequate geographical disaggregation that can reveal geographical variations in phenomena. This involves the creation of a spatial data infrastructure enabling standardized location references for mapping spatial location to statistical data units. This allows statistical data to be referenced to the finest geographical scale possible, down to a geographic coordinate.

661. The assignment of a unique identifier to each location allows linkages between with other statistical and geospatial data associated with the same geographic space. The geocoding of statistical data considerably expands analytical possibilities, including integrating them into indicators and other data, but also analysing the data from a geographical point of view.

662. The provision of these common geographies allows the generation of statistical data in a consistent manner, through cartographic grids or units with administrative or statistical boundaries. These allow statistical data to be aggregated/disaggregated at different levels. Common geographies create common language and allow for integration across domains.

663. Improving availability of more granular or local data may include:

(a) Using new data sources and increasing usefulness of censuses for climate change-related statistics (see also section 9.3 above);

(b) Developing methods for publishing on low aggregate level without confidentiality problems;

(c) Working with cities and municipalities to support them in producing local data in line with agreed standards and definitions.

664. Improvement of availability of geospatially enabled data may include:

(a) Working with the geospatial community on improving interoperability of climate-related statistical and geospatial data;

(b) Consistently georeferencing variables and disseminating geospatially enabled datasets;

(c) Use of GIS tools and platforms for data analysis, integration and dissemination.

665. Improving interoperability of statistical and geospatial data is a complex topic that the statistical community has been working on in various fora for a number of years, for example:

(a) The United Nations Global Statistical Geospatial Framework (UN-GGIM, 2018), which is being implemented in an increasing number of countries, is a fundamental reference for NSOs to be able to respond to the increased demand for better integration of statistics with geospatial information;

(b) INSPIRE programme in the EU;

(c) Since 2020, the Conference of European Statisticians and the UN-GGIM: Europe hold a joint day of their plenary sessions;

[140] https://www.cso.ie/en/statistics/waterandwastewater/domesticmeteredpublicwaterconsumption/

(d) UNECE, Eurostat, UN-GGIM: Europe collaboration on integrating statistical and geospatial data - workshop, Task Force on standards and Survey on the Integration of Statistical and Geospatial Information;.[141]

(e) In the context of specific statistical domains:

  (i) CES Recommendations on measuring hazardous events and disasters;

  (ii) Strategic Framework on Geospatial Information and Services for Disasters;

  (iii) An Inter-agency and Expert Group on SDG Indicators (IAEG-SDGs) subgroup on the use of geospatial data for SDGs;

  (iv) Ecosystem Accounts.

Italy

Register of places. ISTAT is changing its production processes, aiming to develop an Integrated System of Statistical Registers, including the Statistical Register of Places (Ferruzza, A., 2023). The geographical statistical information of Statistical Register of Places has an increasing potential to consider statistical measures related to climate change and sustainability.

The Statistical Register of Places is a complex system with several components. For each register component, variables are being built detailing several characteristics of the entity under study and information on their quality. The challenge is the production of spatial information able to respond to the heightened need for detailed statistical data integrating these different components. The goal is to have a detailed geography for the statistical units of all the other social and economic statistical registers and of some surveys to improve statistical analyses.

The main administrative source is the Cadastral Administrative Archive, which registers any Real Estate unit, and therefore includes not only residential buildings. Other sources are considered: Regional Cartography, National Geoportal, Open Street Maps Buildings have been georeferenced. The geographical center of gravity is used to improve the quality of the georeferencing of the statistical units. Addresses of buildings and dwellings are also considered.

Population grids.[142] ISTAT publishes population grid statistics as an alternative to population statistics for administrative areas. Population grids are a powerful tool to describe the society and to study the interrelationships between human activities and the environment. They are particularly useful for analyzing phenomena and their causes which are independent of administrative boundaries, such as, fires, flooding, commuting and urban sprawl, air pollution.

The Netherlands

Statistics Netherlands uses major sub-national administrative areas, as well as smaller sub-national administrative areas for georeferencing statistical data both for internal use and dissemination. For example, see: GHG emissions (Total, CO_2, N_2O and CH_4) by municipality; see NL-maps.

Statistics Netherlands collaborates with municipalities through Urban Data Centres[143]. The Centres combine strengths of Statistics Netherlands: e.g. regarding data infrastructure, data processing and privacy protection, with municipalities' knowledge of local policy issues to find evidence-based policy solutions and shed light on the local implications of societal developments. The Centres allow local authorities, businesses and institutions improve their understanding of actual developments within the municipality.

Poland

Satellite remote sensing. Statistics Poland uses remote sensing to present more detailed data at low territorial levels on:

(a) Identification of crop species, information on the growth rate of crops and their growth conditions e.g. amount of water in the soil;

[141] https://statswiki.unece.org/display/GeoStat/Integration+of+Geospatial+and+Statistical+Data+Home
[142] https://www.istat.it/it/archivio/155162
[143] https://www.cbs.nl/en-gb/our-services/urban-data-centres

(b) Identification of green areas and determination of the quality of plants in these areas;

(c) Experimental statistics for SDGs (11.3.1 - Ratio of land consumption rate to population growth rate, 11.7.1 - Average share of the built-up area of cities that is open space for public use for all).

The main source of data acquisition is the Copernicus project conducted by the European Space Agency (ESA). Satellite data are often more accurate than data obtained from statistical surveys and administrative sources. The satellite provides data at a high spatial resolution e.g. Sentinel 2 – 10x10 meters. Moreover, earth observation data usage generates lower costs and offers better timeliness.

Satellite data allows for more efficient use of data from administrative sources (e.g. for segmentation of satellite images). In addition, it enables faster acquisition of target data, including acquisition of agricultural and environmental data at a low level of aggregation and in a spatial perspective. In the future, this will reduce the burden on respondents and interviewers by partially excluding topics covered by satellite remote sensing from the surveys. Moreover, implementation of remote sensing-based data will extend Statistics Poland's capabilities in data presentation (especially in creating maps and charts for selected areas).

City statistics. Statistics Poland provides a dedicated tool called City Statistics Application[144] to assist numerous users who need short-term detailed data for highly urbanized areas. The tool allows timely access to automatically generated ready-to-use reports (analyses, charts, tables) shortly after data collection and comprises data on labour market, wages and salaries, industry, construction, housing, internal trade and economic entities. Data are available for 18 regional capitals (representative survey). This is a result of compromise between capabilities (lower levels) and user expectations. A standardized and coherent presentation enables comparison between individual cities. Data are available for the current year, as well as for the two preceding years.

Improved delimitation of territorial units: Delimitation of Rural Areas[145] was developed to provide more detailed information on rural areas. More than 93% of the area of Poland is comprised of rural communes, which are inhabited by 40% of Poland's population. The existing dichotomous division of communes into urban and rural was based on the application of legal criteria, not taking into account their demographic and economic aspects. Therefore, it was necessary to develop a new delimitation for statistical purposes. "Agglomeration" type communes have a close relationship with the core city with their services, workplaces etc. "Non-agglomeration" communes are characterized by low population density. They are often peripheral and have many social and infrastructural problems. A new delimitation of rural areas has been developed to enable better presentation of the specific characteristics of communes and (especially) rural areas. The new delimitation divides rural areas into 2 main groups (agglomeration and non-agglomeration) and high- and low-density subgroups. Such a classification also provides necessary information for deeper analysis of the causes and effects of demographic and socio-economic phenomena, taking into account their spatial diversity.

United Kingdom

Achieving more granular breakdowns is challenging with survey data, in particular because of the limitations caused by sample sizes. However, the United Kingdom Department for Energy and Net Zero does publish the following breakdowns:

(a) Sub-national electricity consumption data for electricity, gas and other fuels

(b) Sub-national total final energy consumption data

(c) Road transport energy consumption in the UK at regional and local authority level

(d) Local authority and regional estimates of GHG emissions

(e) An interactive map of UK Emissions

(f) Mapped emissions from individual sources in the industrial and commercial sector

(g) Emission map data for carbon dioxide as carbon

[144] https://statystykamiast.stat.gov.pl/ (in Polish)

[145] https://stat.gov.pl/en/regional-statistics/classification-of-territorial-units/delimitation-of-rural-areas/

9.5 Collaboration with researchers and academia

666. As discussed already in sections 2.1.2.4 and 2.1.3 (see also Figure 2.6), the scientific community plays a key role in climate change policy and data landscape.

667. Many NSOs already collaborate with researchers on development of statistical methodologies and make their data available in a secure manner for research purposes. Strengthening NSO collaboration with researchers and academia in the climate change domain could also include:

(a) Producing synthetic data

(b) Collaboration with researchers advising governments on the policy assessment

(c) Collaboration on reaching the public, debating on the basis of objective data

(d) Collection of data needs for research purposes

(e) Involvement of NSOs alongside researchers in the development of IPCC methodologies and identifying data sources for IPCC experts

(f) Collaboration on education and training

668. It is also worth mentioning that some statistical offices are also research institute themselves, e.g. ISTAT and STATEC Luxembourg, and collaborate with research departments of other government agencies.

Armenia

In 2023, by the decision of the Presidency of the National Academy of Sciences of the Republic of Armenia, Permanent Professional Problem Councils were established within the Academy to organize discussions on the emerging issues in various fields of science and technology, to form a professional opinion based on scientific and expert analyses, and to advise state government bodies and departments, and private companies. The Councils are formed on issues of investment, special purpose, biodiversity and biosafety, seismology, earthquake-resistant construction, natural and man-made disasters, terminology, historiography and cultural studies.

The Councils include scientists from scientific organizations of the National Academy of Sciences, scientific and educational organizations operating outside the system of the Academy, foreign members of the Academy, and representatives of state authorities of the Republic of Armenia and the private sector. Discussion of a particular public administration, economic, social and environmental problem in the relevant council will ensure its comprehensive professional study and allow, based on analysis, the selection of appropriate strategies to solve the problem in accordance with modern scientific standards. The presence of such platforms, operating on a permanent basis, will also, firstly, increase the level of interaction between government bodies, the private sector and science, and secondly, increase the volume of applied research, determined by demand from the economy. Based on this, state administration bodies and departments were instructed to apply to the relevant Council of the National Academy of Sciences to obtain a professional expert opinion when solving existing problems in the field of their coordination.

9.6 Perceptions, attitudes and behaviours

669. As elaborated in Chapter 7, the United Nations Framework Convention on Climate Change, the Paris Agreement and the 2030 Agenda for Sustainable Development all recognize that their goals cannot be achieved without involving all levels and sectors of government, the private sector, civil society and the public. The IPCC Sixth Assessment Report states that the systemic change required to achieve rapid and deep emissions reductions and transformative adaptation to climate change is unprecedented in terms of scale, and systems transitions required to achieve it include reducing and changing demand through, among others, socio-cultural and behavioural changes (IPCC, 2023a). The perceptions and attitudes of citizens related to climate change can therefore be considered as an important enabler of climate action and people's behaviours and a key component in mitigation, adaptation and just transition. Understanding these three elements can provide policymakers with key insights for policy design and monitoring.

Italy

ISTAT conducts an annual Multipurpose Survey "Aspects of daily life" on citizens' habits and the quality of life, both at the individual and household level. The survey covers citizens' opinions about environmental problems and eco-friendly behaviours (Facioni, C., M. Romano and A. Villa, 2023).

The Netherlands

Statistics Netherlands produces statistics on perceptions and behaviours related to climate change, for example, regarding the share of Dutch people who are concerned about the impact of climate change[146] and the share of Dutch drivers who sometimes deliberately leave their car at home.[147]

This Perceptions Survey was, again, conducted from 7 February to 21 April 2023, covering such issues as sustainable housing, sustainable mobility, sustainable food, and climate-conscious lifestyles. Over eighteen-thousand adults took part in the survey. Two news releases were published based on the results.[148] The complete web publication[149] with details on the design and implementation of the study is available in Dutch only.

Switzerland

The Swiss Federal Statistical Office conducts a survey on environmental quality and behaviour.[150]

United Kingdom

ONS runs regular (fortnightly) surveys of both the public (via Opinions and Lifestyle Survey) and businesses (via the Business Insights and Conditions Survey). These enable ONS to capture attitudes and actions on climate change along with many other topics. Relevant questions from both surveys are reported in ONS publications, e.g. a recent survey asking about climate change and the recent heatwave.[151]

---

[146] News release from 9 June 2021: https://www.cbs.nl/en-gb/news/2021/22/three-quarters-of-dutch-concerned-about-impact-of-climate-change

[147] News release from 9 June 2021: https://www.cbs.nl/en-gb/news/2021/22/nearly-8-in-10-drivers-sometimes-deliberately-leave-their-car-at-home

[148] News release from 28 November 2023 https://www.cbs.nl/en-gb/news/2023/48/three-quarters-of-dutch-adults-worry-about-impact-of-climate-change and from 29 November 2023: https://www.cbs.nl/en-gb/news/2023/48/more-dutch-citizens-in-favour-of-nuclear-energy

[149] https://longreads.cbs.nl/klimaatverandering-en-energietransitie-2023/

[150] https://www.bfs.admin.ch/bfs/en/home/statistics/territory-environment.assetdetail.11527958.html

[151] Public opinions and social trends, Great Britain - Office for National Statistics: https://www.ons.gov.uk/peoplepopulationandcommunity/wellbeing/bulletins/publicopinionsandsocialtrendsgreatbritain/3to14august2022

Chapter 10 Conclusions, recommendations and future work

10.1 Conclusions

10.1.1 General conclusions

670. Climate change is one of the major societal issues. It is a global phenomenon that affects everyone well into the future and, therefore, receives a lot of attention. The 2015 Paris Agreement was a milestone that established new global goals on this topic. National and local climate objectives generally all stem from this agreement. Governments, organizations, industries, and academia involved in reaching these goals - the climate change community - define the key topics of interest around climate change. In the context of producing and using the data/information needs, they form the demand side. The statistical system is on the supply side. Although NSOs are not the main players, they have a lot to contribute.

671. This report describes the demand and supply of data relevant to achieving climate change goals in the form of a compendium - starting from the concepts, definitions and policy context, through actors and their data demands, to everything the statistical system has to offer on each of the themes discussed. The theory is interlaced with practical examples from countries. In principle, a lot of data is already available and more could be produced. Different priorities, levels of statistical development, engagement with the climate change community, and available resources are inhibiting progress. Data from many domains are needed, and significant gaps between demand and supply persist. NSOs could play a bigger role in addressing these gaps. This would require strengthening primary data collection, engaging the communities, and building capacity at the national level.

672. The statistical system is more than NSOs and does more than just produce data. To be able to provide the right and desired information, the statistical system should actively map and track user needs and experiences. Data can also be produced by organizations other than NSOs. Here, the role of the data steward overseeing and ensuring the quality and suitability of the system's data assets, including the metadata for those data assets, would be beneficial.

673. An essential point is that no one is expecting the NSOs to do everything. In the broad field of climate change-related data, no one can do it all. It is rather about establishing collaborations, sharing the work, and playing a role according to the institutions' expertise. These collaborations should be governed by governments, involve data institutes and academia, and be fueled by input from industry associations.

674. There is a need for capacity building on both the demand and supply sides. The complex institutional landscape and diversity of inter-related subjects make it difficult to keep an overview while connecting the technical details to the bigger picture. Capacity building has to account for the differences in the maturity of various statistical systems in both developing and developed countries. It should also pay attention to the demand side to better understand what the statistical system can offer or already offers.

675. The main conclusion is that a large system is being built around achieving climate goals; however, the statistical system is lagging and unable to play the role it has for other societal issues. National statistical systems have mechanisms in place that allow NSOs to address new topics, include them in the statistical programme, establish a budget, and coordinate with other data providers. These mechanisms seem to work better for the traditional fields of work of NSOs related to social and economic statistics. In the case of environmental data, parts of the usual institutional infrastructure seem to be missing. At the national level, this could be due to the lack of environmental divisions in NSOs and weaker connections to ministries, knowledge institutes, and other data holders. At the international level, institutions that are central to international or regional statistical systems are not very engaged in the climate change community. Statistical frameworks developed by the statistical

system without close engagement with users can end up inward-looking and using definitions that differ from those for climate objectives. Further, the organizations working closely with NSOs are not always active players in the climate/environment field and the active international organizations that are, may not be as involved in the work with NSOs.

676. The report was produced by a dedicated Task Force within approximately a two-year time frame. The Task Force members have learned a lot about the role of NSOs and related to the procedures and information needs of various stakeholders by working together. The collaboration has allowed the team to improve their skills and be exposed to different perspectives and expertise. For them, the result of the Task Force's work is the starting point for proceeding with their own work in their respective organizations.

677. The following sections present the condensed conclusions from each chapter.

10.1.2 Institutional landscape and the role of NSOs

678. NSOs are independent, professional providers of data on key societal topics. They serve governments, economic actors, and the public with official statistics and services related to data. They are the leaders of national statistical systems with mandates for and expertise in data collection, safeguarding, processing, and dissemination. NSOs work in line with the Fundamental Principles of Official Statistics; one of these Principles is to strive for international comparability.

679. Many NSOs are currently discussing how best to position themselves in the constantly expanding data ecosystem to protect the integrity and relevance of official statistics and contribute to the treatment of data as a long-term asset that benefits society. In numerous countries, the role of NSOs is already evolving, leading to NSOs undertaking data stewardship roles, which may include:

(a) Contributing to the development and promotion of data strategies, policies, and principles;

(b) Coordinating the standardization and harmonization process and supporting interoperability;

(c) Providing expertise in data quality and developing and promoting data quality frameworks;

(d) Contributing to improving data literacy;

(e) Providing expertise and services related to methods, tools, and capabilities;

(f) Building partnerships across the public sector.

Figure 10.1.
Actors in the institutional landscape

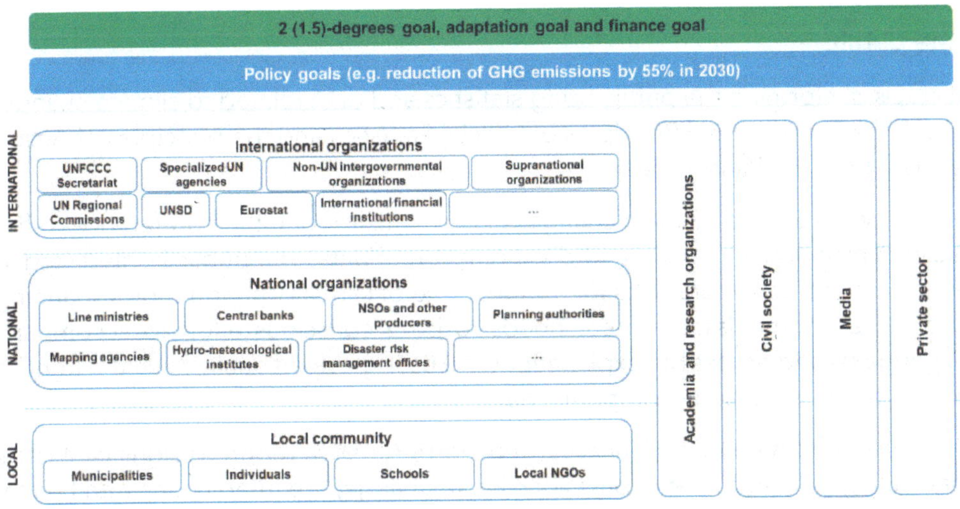

Source: Authors.

680. In the climate change domain, there is a lot of emphasis on the relationship between policy and science. This report attempts to incorporate the third category, statistics and data, which are the

foundation of both sound decision-making and scientific research. Actors representing all three categories at international, national, and local levels create a complicated and extensive institutional landscape (see Figure 10.1).

681. The actors in that landscape may play various roles in the context of climate change-related data, from data holders to statistics producers to data users (Figure 10.2). For example, NSOs will, in most cases, be data compilers, but in the context of GHG inventories, they can be seen as a data holder. A national energy agency can take the role of a data holder as well as data user.

Figure 10.2.
Roles in producing climate change-related data

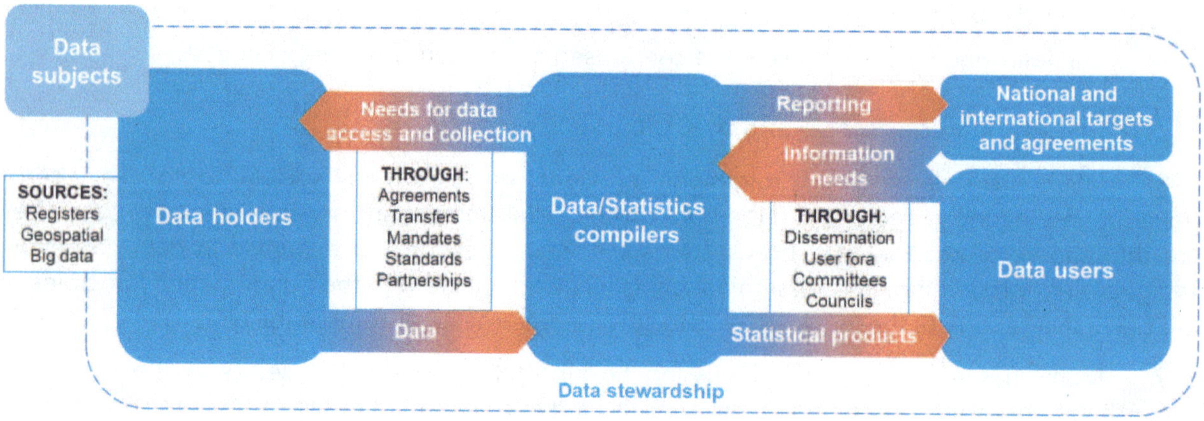

Source: Authors.

682. As shown in the report, many statistical frameworks and indicator sets for climate change-related statistics are already in place: the CES Recommendations on Climate Change-Related Statistics, Framework for the Development of Environment Statistics, System of Environmental-Economic Accounting, International Recommendations for Energy Statistics, Disaster-related Statistics Framework, the CES Set of Core Climate Change-Related Statistics and Indicators, CES Set of Core Disaster-Risk-Related Indicators, the Global Set of Climate Change Statistics and Indicators, Sendai Framework indicators, and the SDG Global Indicator Framework. These provide a starting point, a toolbox and a structure for producing statistics and data useful for policy applications, which are presented in the following chapters. However, it is worth noting that the overview of relevant frameworks is not exhaustive, as examples from almost all statistical domains can be found throughout the document.

10.1.3 Informing the public

683. A key role of NSOs is informing the public using statistics and data related to climate change. This role aligns with both the Fundamental Principles of Official Statistics and the principles of Action for Climate Empowerment from the Paris Agreement.

684. There are several user groups for climate change-related statistics, and their preferences differ in terms of communication products, channels, and dissemination of data. The public is interested in both in-depth stories and insight-driven graphics for a wide range of climate change topics spanning environmental, social, and economic issues. There needs to be a conscious effort to accompany statistics with both expert-level and accessible explanations. All user groups equally have the right to know and make informed decisions based on impartial data.

685. Interviewed journalists from Europe, Caucasus, and North America reported significant gaps in data availability, particularly related to the impacts of climate change, the interactions with health, the economy, and nature, and adaptation. There are also broader difficulties with the level of granularity, timeliness, and accessibility of data. Finally, they made several recommendations to NSOs.

686. Country examples provide insights on how the NSOs can work to disseminate and communicate climate change-related statistics, actively reach out to the public, use social media and other channels, work on machine readability and searchability of statistics, and coordination between statistical experts and communications staff within the NSO.

10.1.4 Reporting under the Paris Agreement

687. The Paris Agreement is built around national climate targets (nationally determined contributions), plans to achieve them, and tracking and reporting of countries' progress towards the implementation and achievement of these plans conducted under the enhanced transparency framework (ETF). The ETF will serve as the primary vehicle for the governments to demonstrate progress in implementation and achievement of their NDCs. It needs to be prepared and communicated every five years. Also every five years, governments will come together to participate in the global stocktake – a process of periodically taking stock of the implementation of the Paris Agreement to assess the collective progress towards achieving the long-term goals of the Agreement.

688. The ETF is applicable to all Parties and builds upon the existing measurement, reporting, and verification system under the Convention. For non-Annex I Parties, the level of detail in the reporting until now has been substantially lower and although some developing countries have good inventories, the learning curve for many may be steeper. Meeting the reporting obligations under the Paris Agreement is a process, and the establishment of robust and sustainable institutional arrangements is a key enabling factor. This includes building capacity and expertise within a country so that necessary information can be collected and provided in a timely and reliable manner. The UNFCCC/IPCC guidance on national institutional arrangements is not prescriptive, and countries are free to decide on the institutional setting that allows the country to report and deliver sustained quality improvements in the reporting while considering their national circumstances.

689. Key elements of a robust national system include:

(a) Formalization of the process legally, including data arrangements;

(b) Formalization of roles and responsibilities of different actors;

(c) Internalization of knowledge and improving capacities;

(d) Closer collaboration between experts;

(e) Addressing lower-level data requirements to improve the national or regional inventory;

(f) Increased policy relevance, public awareness and scrutiny;

(g) UNFCCC reviews and audits;

(h) 'Slowly but surely': a stepwise approach to sustained improvements.

690. NSOs have a role to play in the ETF, even if the exact role and degree of involvement may differ by country. At minimum, NSOs are responsible for producing a significant share of the input data to GHG inventories, such as statistics on energy, agriculture, waste and industrial production. NSOs can also be formally involved in the reporting, e.g. in data quality assurance or by providing more detailed data to the compilers in a secure manner. In a few countries, NSOs are responsible for at least part of GHG emissions estimation or inventory compilation. In any case, NSOs should be actively involved to understand how data are used, discuss quality issues, and improve data collection if needed. Furthermore, NSOs produce a lot of statistics that can be used to inform or monitor actions needed to implement and track the NDCs, which countries submit under the Paris Agreement.

691. Improving the availability of internationally comparable data will also facilitate the subsequent collective assessments of progress for the Global Stocktake.

10.1.5 Statistics for climate change mitigation policymaking

692. Mitigation of climate change is defined as human intervention to reduce emissions or enhance the sinks of GHGs. In climate policy, mitigation measures are technologies, processes, or practices that contribute to mitigation, such as renewable energy technologies, waste reduction processes, and public transport commuting practices.

693. In the report, the demand side is illustrated by national and subnational policies and actions, sourced from the OECD Inclusive Forum on Carbon Mitigation Approaches and the nationally determined contributions; data needs are related to the Energy Supply, Transport, Buildings, Industry, Agriculture, LULUCF, Waste sectors, and cut across the sectors.

694. The supply side includes a variety of well-chosen statistics on GHG emissions and directly related topics. The chapter also highlights that informing and monitoring mitigation policies should not only focus on GHG emissions but also cover the enablers essential for achieving a successful transition(s). In addition, the importance of energy statistics needs to be emphasized, just like other official statistics used as essential inputs for inventory compilations (as activity data) and other statistical data relevant to achieving a successful transition.

695. It is overwhelming to see what data is already available on both the demand side (Section 4.2) and the supply side (Section 4.3). Within this report, it was not possible to fully confront the two sides to expose the data gaps, which would require more time and specific, detailed knowledge, which the authors (and the small networks around them) did not have. Using global expertise and a step-by-step approach would be needed to reveal all the data gaps, sector by sector, theme by theme, and topic by topic. To this end, appropriate meeting and discussion platforms can be set up at the national level, coordinated by NSOs, and at the international level.

696. Overall, two main roles for NSOs are distinguished. The first is focused on collecting primary data that are relevant for climate-change indicators and providing activity data for the inventory compilation process. In the second role, NSOs can provide tailor-made data to track the progress of mitigation policy actions. However, more efforts, resources, and capacity building are required to close the gap between the lower level (e.g. primary energy data) and the higher levels (e.g. energy-related climate indicators for policymaking). For both purposes, energy statistics play a central role.

697. For both NSO roles, the compilation and provision of official statistics facilitate the monitoring of the mitigation domain in the Enhanced Transparency Framework under the Paris Agreement. To that end, NSOs can set up dashboards to disseminate harmonized indicators. One of the challenges that NSOs face is the alignment of concepts, methods, and classifications used under the Enhanced Transparency Framework with those used in official statistics.

10.1.6 Statistics for climate change adaptation policymaking

698. Climate change adaptation can be understood as the process of minimizing adverse impacts of climate change by reducing exposure and vulnerability and increasing adaptive capacity. From an NSOs perspective, informing adaptation policymaking does not exclusively involve producing data and statistics on the adaptation measures alone - the information about the adverse consequences is just as important.

699. Adaptation responses require information on not just the **hazards** themselves but also the level of **exposure** to them in a given country, the **vulnerabilities** of the population and economy, and the **capacity to cope** with hazardous events. Together, these elements determine the level of **risk** of adverse consequences of climate change and the human response to climate change in a society. The adaptive measures are changes in processes, practices, and structures intended to reduce the levels of climate change-related risk through the elements of exposure and vulnerability, thus lessening the final adverse impacts. Figure 10.3 illustrates how mitigation and adaptation actions aim to reduce climate risks and impacts.

Figure 10.3.
How adaptation and mitigation actions can reduce climate-related risks and impacts

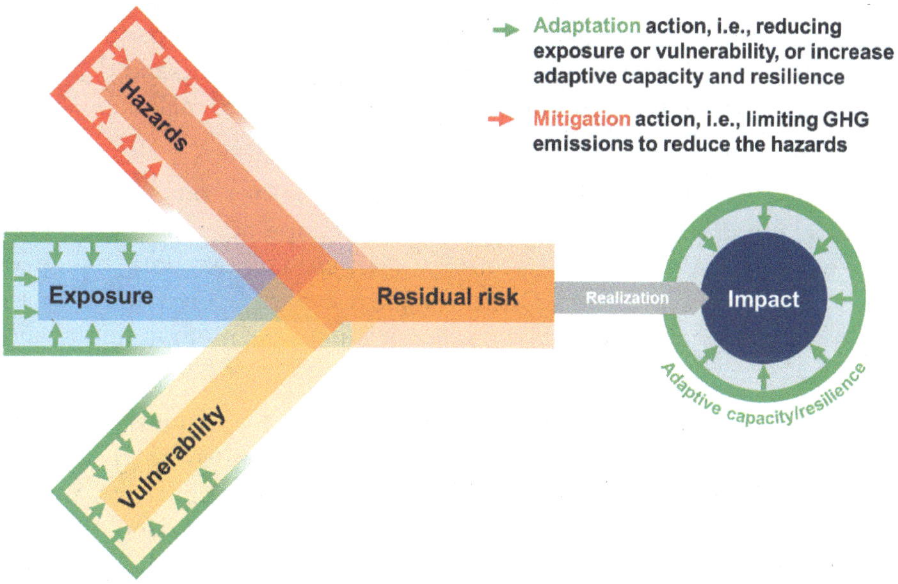

Source: Authors.

700. To produce data required to adequately inform adaptation responses, NSOs should understand the scope of information needed for effective climate change adaptation policymaking. A useful tool for achieving this is the NAP process, which enables identifying country's adaptation needs and creating strategies and programmes to address these needs.

701. The 2022 Synthesis report for the technical assessment component of the first global stocktake, State of adaptation efforts, experiences and priorities mentions the key challenges related to adaptation faced by countries, including data availability and information, and insufficient coordination among the main stakeholders that hold climate data and thus have a role to play in helping societies adapt to climate change. To identify key stakeholders, NSOs could evaluate them based on criteria such as the ability of the stakeholder to identify, produce, and share data that is relevant to the country's interests and decision-making and reporting processes.

702. Central to data and statistics related to climate change adaptation is the conceptual and statistical understanding of climate-related hazards in terms of their occurrences, exposure, vulnerability of, and eventual impacts to population and ecosystems. In this regard, the **disaster risk reduction** community has long dealt with these measurement and data collection issues pertaining to extreme events and disasters, including climate-related hazards.

703. Hence, the climate change community, in its effort to support informed decisions on climate change adaptation, would benefit from engaging the disaster risk reduction community to build on existing knowledge and expertise to promote greater alignment in the production and use of climate change and disaster-related statistics. Subject to specific national contexts, NSOs could play a key role in facilitating closer collaboration between the climate change and disaster risk communities both at the national and global levels.

704. Country examples help to illustrate that there is no 'one-size-fits-all-solution' that can apply to the unique context of each community, business, organization, country or region. NSOs can contribute in various ways depending on what is of most relevance in each specific context e.g., through applying and improving activities already undertaken by NSOs on other topics to climate change adaptation, such as coordination with other organizations, producing relevant statistics across many statistical domains, linking data from different sources, focusing on the need for enhanced geospatial data, etc.

10.1.7 Statistics for just transition policymaking

705. Just transition is an economy-wide process that produces the plans, policies, and investments that lead to a future where jobs are green and decent, emissions are at net zero, poverty is eradicated, and communities are thriving and resilient. Ensuring this requires identifying and describing (socially) vulnerable groups who may suffer the most from transition risks related to mitigation, adaptation, and other climate change-related policies.

706. NSOs are well-positioned to measure distributional effects. They can provide timely, accurate, and granular (micro)data, not only on socioeconomic and sociodemographic aspects but also on living conditions. To achieve proper (micro)data linking, NSOs can facilitate data sharing, in which the confidentiality of individual data is respected. One of the main roles of NSOs is in harmonizing regular data compilation is establishing standards regarding definitions, methodology, and indicator development.

707. An inventory of best practices is needed to help integrate a just transition perspective into NSOs' work on climate change-related data and statistics, including how to engage with stakeholders, how to incorporate social equity and inclusion considerations into data collection and analysis, and how to ensure that data is used effectively to support policy and decision-making.

10.1.8 Climate-related costs, transfers and financial flows

708. Climate finance is currently a concept without an agreed-upon international definition but is an enabler of climate change mitigation, adaptation, and just transition. Under the Paris Agreement, there is a separate goal on making financial flows consistent with a pathway towards low GHG emissions and climate-resilient development as well as a requirement for developed countries to provide resources to assist developing countries. At the same time, the financial dimension of climate change is broader. Investors consider companies′ positions vis-a-vis climate change and sustainability objectives (energy use, waste management, pollution), which has led to an increase in environmental, social, and governance (ESG) reporting. Climate change has also become a key concern for international regulators and an important factor for the stability of national and international financial institutions.

709. There are a number of issues related to the financial aspects of climate change that should be of interest to the official statistics community. These range from defining, tracking, and measuring climate finance through differentiating between climate, green, and sustainable finance and describing the information needs of the financial sector.

710. Many initiatives are already in place in this thematic area, from formal bodies under the UNFCCC to other international initiatives such as DGI-3 and task forces in the central bank/financial sector community. At the same time, many gaps and challenges persist, and many recommendations about the need for further work have been made.

711. Potential areas where NSOs can contribute include coordination/cooperation, statistics on government expenditures, subsidies and transfers, supporting national climate finance reporting, measuring climate impacts of economic and financial activities, producing data for assessing physical and transition risks, supporting efforts towards standardization, and measuring climate investment. A contribution of statistical offices is needed to solve some of them, while for others, it may not be necessary for the NSOs to be actively contributing while still keeping a strategic and coordinating perspective. SEEA framework can play an important role in bringing the relevant information into a common and consistent framework. The ongoing revisions of the SEEA-CF, GFS and COFOG provide the opportunity to make these frameworks even more relevant for climate change. The topic is vast, which is why different actors can lead on specific issues. Responsibilities for these statistics vary from country to country, and the role of the NSO can vary accordingly. Regardless of the institutional arrangements in place, it should be a collaborative effort.

10.2 Recommendations

712. The statistical community already provides a wealth of valuable data that can be used to answer questions related to climate change policymaking. In order to play a proactive and even bigger role in this space, statisticians need to ensure a solid understanding of climate change science and policy. Several recommendations are proposed below that aim to ensure the production of innovative and insightful data products that can help achieve climate goals.

713. The **first overarching recommendation** for all NSOs to move forward in this domain is to get engaged with the climate change community. Through such engagement, NSOs could better understand the data needs, and the climate change community would better understand how statistical systems work and their strengths. As policymakers think of measures within an administrative-legal framework, data producers think in terms of a statistical-scientific framework. One perspective focuses on goals and policy measures, while the other aims to harmonize definitions and implement a monitoring programme. Progress is necessary to increase the engagement between the statistical and climate change community, instilling high-level commitment, collaboration in the workplace, and a high degree of reciprocity.

714. The **second overarching recommendation** for all NSOs is to start reviewing existing data collections and adjust to user demands where possible. Particularly for NSOs that have not worked on this topic, it is advised to start with statistics and data that are already available, then build up a user community and start conversations on how the information currently produced by the NSO can support climate change policies. Existing tools, such as the Climate Change Statistics and Indicators Self-Assessment Tool (CISAT), can be used for this purpose. The next step is to begin identifying gaps in information for climate action and producing new relevant statistics where possible, preferably working together with the climate change community.

715. The scope of climate change issues is colossal if we take into account all actions that nearly all countries in the world want to undertake. Equally broad is the scope of climate change-related statistics as "*Environmental, social and economic data that measure the human causes of climate change, the impacts of climate change on human and natural systems, the efforts of humans to avoid the consequences as well as their efforts to adapt to the consequences*" (UNECE, 2014). Now is the time for NSOs to reinvent themselves and respond to this major social challenge. The following more detailed recommendations are divided into relationship management, process reinforcement, and content development and will support the implementation of the overarching recommendations. Priority should be given to the development of climate change-related statistics in all three aspects.

10.2.1 Cooperation and capacity building

716. **Collaboration** is key. This is clear considering the variation in related domains and the multitude of task forces, working groups, and committees set up under various international organizations such as UNFCCC, UNSD, UN regional commissions and other UN agencies, OECD, Eurostat, etc. For this, engagement and dialogue should be established between data producers and users, in particular policymakers; ways to do so must be explored. An option is to meet at the COP, as the climate change community currently meets there. Additionally, collaboration with researchers and academia on methods must be strengthened; IPCC is critical in this capacity. It is likely that countries have researchers in academia, outside of government policy and advisory roles, working in various climate change-related research programmes. This collaboration could focus on data gaps and the use of data for research.

717. Within the NSOs themselves, it is more about organization than collaboration. The main recommendation for NSOs here is to ensure that there is a **higher-level, strategically thinking manager** explicitly responsible for climate change and environment domains; within a separate environment/climate change statistics division for example. Currently, statistical offices are usually organized around social and economic topics. Lack of higher-level management attention hinders

both development within the NSOs and maintaining relationships with external partners. In addition, it is also relevant to organize coordination between experts from various domains within NSOs.

718. Expand the opportunities for **capacity building** both for explaining the policy framework and further development and extensions of official statistics and other tailor-made (micro)data. Priority should be given to the variety of data needed to monitor the clean energy transition and other transitions envisaged in NDCs. Further topics are data stewardship, microdata linking, and consistent data compilation from different perspectives.

719. In the context of **reporting under the Paris Agreement**, a good practice is to have joint working groups between, for example, inventory compilers and those responsible for energy statistics at the national level. This type of activity improves the quality of the energy data in the inventories and the energy data reported by the statistical offices, as well as the coherence between both data sets and the cooperation between the different groups.

720. To build capacity related to the data used in the climate reporting process, NSO staff can also be trained as UNFCCC reviewers, if appropriate and possible in national circumstances. This would be beneficial for addressing data gaps and improving data quality, allowing NSO staff to learn about the climate reporting process, which is not easy to understand without being involved in practice, and establishing relationships with the climate community. However, the first priority of NSOs should remain to produce high-quality statistics.

721. Concerning **climate change adaptation**, NSOs should collaborate more closely with the disaster risk management community and producers of disaster-related statistics. To understand the scope of information required for effective adaptation policymaking, NSOs should inform themselves on the scope and progress of their country's NAP process.

722. Regarding the **financial aspects**, it is recommended that a national coordination committee, including the NSO, be established to facilitate cooperation and avoid any overlaps to best leverage the work that NSOs already do.

10.2.2 Governance and procedures

723. Many governance and coordination mechanisms in relation to climate change goals are already in place. The involvement of data and statistics compilers and data holders in these mechanisms is highly recommended. It is also recommended that statisticians become active and involved in existing mechanisms than creating separate mechanisms for statisticians alone.

724. Focused international meetings are to be established, possibly with national counterparts, bridging the data gap between what NSOs can contribute (supply side) to what is needed in the mitigation, adaptation, or just transition domains (demand side). The exact meeting format (e.g. one umbrella group with separate discussion groups per sector) depends on the specific mandates of the relevant (inter)national organizations. At the national level, the NSO can play a coordinating role. The presence of data users is key to the success of such meetings. Therefore, it is recommended to hold meetings where users are already present, such as back-to-back meetings.

725. Monitoring and evaluation programmes should be set up to fully harness the power of the statistical system, which works best when a joint programme is agreed upon between agencies and planning authorities. In such a programme, the NSO can work out definitions and best practices for producing data and information. Data collection ideally goes hand in hand with evaluation, ex-post and ex-ante, by agencies and planning authorities. Scenario models used for these evaluations follow the same definitions statisticians use to record the occurrence of the same phenomena so that the past, present, and future are in line with each other.

726. NSOs should consider taking on a data stewardship role in the context of climate change-related statistics and data to advance collaboration and use of data assets. Such a role is particularly relevant in this domain due to the complicated landscape with many data producers, compilers, and users, as mentioned at the beginning of this chapter.

727. Regarding **climate change mitigation**, a closer dialogue between NSOs and Enhanced Transparency Framework (ETF) stakeholders should result in more detailed descriptions of the mitigation policies data needs, more effective collaboration in solving identified data gaps and more guidelines on how the requested statistical data should be produced. Existing tools, such as CISAT, can be used in the process.

728. For **climate change adaptation**, NSOs should coordinate with other key stakeholders and establish their roles and information needs. The key is to prioritize user demands to determine which adaptation statistics are of most relevance to the context of each country. It is recommended to use tools and frameworks that are readily available, for example, the CISAT tool or by mapping statistical products to adaptation-related concepts illustrated in Figure 5.1. Where possible, NSOs should become involved in the NAP process and, internationally, in processes related to the development of indicators and reporting under the Global Goal on Adaptation. Although currently, there are no groups where statistical offices are actively involved, this is an objective to pursue both nationally and internationally.

729. NSOs should follow the guidance included in the 2023 COP28 decision on reporting for the thematic targets under the UAE Framework for Global Climate Resilience as there is an opportunity to play an active role in the reporting process. The indicators for measuring progress achieved towards these "area" targets (on water, food, health, ecosystems, infrastructure and human settlements, poverty eradication, and cultural heritage) and "process" targets (on the adaptation cycle) are due to be elaborated upon in the 2023 - 2025 UAE - Belém programme, carried out jointly by the Subsidiary Body for Scientific and Technological Advice and the Subsidiary Body for Implementation.

730. In the case of **climate-related costs, transfers and financial flows**, a key area that should be exploited in the short term is facilitating data access primarily for statistical compilation purposes, e.g. by the central banks, but for research as well. This might include locational data and business registers.

10.2.3 Content and development of new information

731. Many areas within the climate change statistics domain would benefit from strengthened data collection. Traditional data collection, such as censuses and surveys, should be examined from the perspective of meeting climate change-related data needs, including addressing information needs related to gender, disability and indigenous populations. New data sources, such as administrative data, earth observations, or electric meters, can complement traditional methods to increase the efficiency of data production or enable the production of new statistics and indicators. Both of these approaches require additional financial resources, expertise, legal basis, IT infrastructure, user demands, and justification for new data collection and high-level support.

732. Availability of **granular, localized, and geospatially enabled statistical data** needs to be improved, through georeferencing all available micro and aggregated data for territorial analysis and adding different layers of information depending on the statistical domain.

733. **Linking environmental data** at the individual level with social and economic data is a unique function that only NSOs can undertake and should be leveraged to address complex policy questions in this domain. With necessary measures and within existing regulation, datasets, including linked microdata, should be made available to researchers who can make full use of their analytical potential.

734. Further, NSOs can play a key role in **establishing and harmonizing definitions**.

735. On **informing the public**, a general recommendation to the climate change community is to make use of the strong and existing communication network of NSOs to inform both the public and specialists. Other recommendations are to:

(a) Map and evaluate user needs and experiences regularly to ensure and contribute to public awareness and public access to climate change-related statistics. Tools like focus groups, analysis of website use, and feedback surveys can be used. Involvement of the communication experts from the NSO is recommended;

(b) Assist users by consolidating links or resources to relevant sources such as national legislation, temperature and weather data, or international comparisons. This can be done through thematic websites or dashboards for climate-related statistics and indicators. Data content can be presented according to the structure drivers-emissions-impacts-mitigation-adaptation. International comparisons and information available in English increase the relevance of such channels for some users.

(c) New and emerging user groups and technologies underline the need for NSOs to communicate widely and proactively, as well as improve machine readability and searchability of statistical products and websites. Service desks and 'under embargo' releases help journalists do their work and make better use of climate change data, indicators, and reports.

(d) Provide user guidance on the different methods/approaches for emissions compilation, ideally both for generally interested users in a short format (e.g. key figures or short fact box) as well as for expert analysts via technical notes and documentation.

736. On **mitigation**, the main recommendation for NSOs is to focus on providing and disseminating statistics and tailor-made data to track the progress of mitigation policy measures as described in the Biennial Transparency Reports. Developed countries' good practices in data provision, data gap filling, and data validation can be shared with developing countries and vice versa.

737. On **adaptation**, in addition to reviewing existing data collections and exploring new ones, it is recommended to provide data both on adaptive measures for reducing climate change-related risks and on the elements that feed into these risks (i.e. climate-related hazards, exposures, vulnerability. and coping capacity), as identified in the NAPs. Although many statistical domains can be relevant for adaptation, in line with the COP28 decision on the Global Goal on Adaptation, key sectors are water, food, health, ecosystems, infrastructure and human settlements, poverty eradication, and cultural heritage. Particularly important in all sectors are also local, granular, and georeferenced data.

738. On **just transition**, the main recommendation for NSOs is to develop harmonized indicators to measure progress towards a just transition, for example on energy poverty or on skills needed for the green and decent jobs that will be available after the transition. A lot of regular and disaggregated data already exists today (for example statistics on demographics, environment, living conditions, consumption, energy use, income, labour, and gender), but there is a need for better cross-analysis and data sharing. An inventory of best practices is needed, including how to engage with stakeholders, how to incorporate social equity and inclusion considerations into data collection and analysis, and how to ensure that data is used effectively to support policy and decision-making.

10.3 Further work

739. To support the implementation of the recommendations proposed above, the following further work at the national and international level should be considered.

740. An important recommendation from this report is **building relations with the user community.** These relations need both active collaboration and institutional support. To establish and maintain these relationships, it is necessary to follow the policy developments, decisions from meetings of the COP, identify formal and informal opportunities for engagement, and keep communicating concretely what official statistics have to offer to the climate change community. This requires dedicated efforts and resources and should be considered an activity on its own. Moreover, the progress in the engagement of the statistical and the climate change communities should be monitored and followed up on.

741. Every country will have its specific institutional setting. Figure 10.4 presents a generic institutional landscape supplemented with symbolic connection arrows. National and international statistical organizations should examine the existing (or not) connections in more detail and undertake further work establishing relationships, initiating collaborations, attending meetings to build the

missing connections that are needed to provide a solid institutional framework for the development of climate change-related statistics.

Figure 10.4.
Actors in the institutional landscape with relations

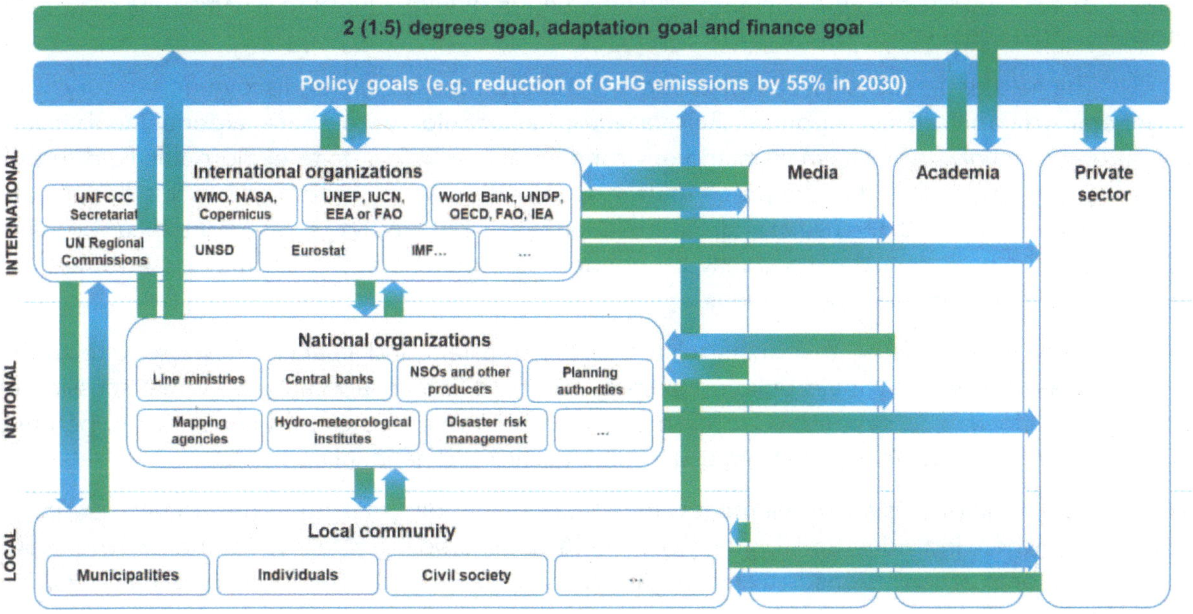

Source: Authors.

742. Based on the work of the Task Force, two areas seem to be of particular interest. First, identifying ways of strengthening the relationship with academia and researchers to improve the flow of information about data availability and gaps, collaborate on addressing methodological issues and development of indicators. The second area is the work of international organizations active in climate change-related statistics – the coordination among them, the extent to which they work with NSOs, and how actively they are involved in the climate community. This should be the first step towards identifying and fully leveraging the avenues for NSOs to engage with the climate change community internationally.

743. As per the Paris Agreement, 2024 marks the beginning of the new reporting cycle. Support for non-Annex 1 countries, which will report for the first time, promotion of involvement of NSOs in institutional arrangements, and use of official statistics in the reporting for all countries are needed. A collection of best practices in reporting and examples of the division of responsibilities between institutions in various countries with different types of national statistical systems will be helpful. Capacity building in relation to the reporting could be organized with UNFCCC for data providers from NSOs, inventory compilers, and data users from international organizations like EEA or UNEP.

744. To continue improving the availability of internationally comparable data, the promotion and support for the primary data collection and the implementation of internationally agreed indicator sets should continue. At the same time, promoting innovative approaches spearheading the development of new statistics is also needed.

745. As the importance of climate change, the scope of topics, and new information demands continue to grow, expert groups in specific domains are already starting to work on climate-related questions within their domains (e.g. gender, poverty, and economic statistics). This distribution of work should accelerate addressing methodological issues and make statistics from all domains more useful for climate change policy and analyses. NSOs and international statistical organizations need to support mainstreaming climate change considerations in various statistical domains while coordinating efforts between relevant expert groups and sharing progress within the bigger climate change statistics community.

746. One established but critically important and promising area of interest is integrating statistical and geospatial data, such as georeferencing statistical data and incorporating the use of geospatial data and remote sensing. Many initiatives are currently in place to advance this aspect of the modernization of official statistics. Thematic sessions could be periodically included in the UNECE Expert Fora for Producers and Users of Climate Change-Related Statistics to keep abreast of the progress and discuss concrete use cases.

747. The field of perceptions, attitudes, and behaviours is a relatively undiscovered territory, but it is generally recognized as important. Further work here could be to share experiences between countries and involving policy makers and environmental psychologists to explore this field further. This topic also is also linked to the problem of misinformation, which is common in this domain and expected to increase with the rise of artificial intelligence's popularity. Moreover, more focus on taking into account cultural considerations may be needed as a generic approach may be insufficient in jurisdictions with indigenous populations.

748. Many topics related to climate-related costs, transfers and financial flows require further international work, which should build on existing initiatives and collaborations. More discussion on how to move forward is needed. A decision about what to prioritize should consider the potential speed of work and involve experts from economic statistics and national accounts.

749. Data availability is a core element within data-driven policymaking. A new idea to explore is data spaces, where individual and aggregated data can be shared in a safe way. Further work for NSOs and international organizations is to make data spaces practical, usable tools for climate change-related data.

750. In order to promote and make use of this document, the content, in particular country examples, should be published in an interactive and easy-to-navigate online format. This could be a live publication, which could be updated as practices develop. It is also encouraged that the report or its key parts be translated into national languages to improve understanding of basic concepts and approaches. The usefulness of sharing Python scripts in GitHub could also be explored as it would be helpful for research software engineers in the domain.

751. Additional materials based on the report could be further developed to facilitate the use and communication of key messages about the role of NSOs to various audiences (e.g. policymakers, statisticians, academia, and journalists).

752. Last but not least, future work should further develop the climate change statistics community and its capacity through sharing experiences, lessons learnt, and knowledge about climate change.

Annex 1. Terms of reference for the Task Force and guiding principles for drafting the report

TERMS OF REFERENCE FOR A TASK FORCE ON THE ROLE OF NATIONAL STATISTICAL OFFICES IN ACHIEVING NATIONAL CLIMATE OBJECTIVES

This document presents the terms of reference for the Task Force on the role of national statistical offices (NSOs) in achieving national climate objectives, which is a follow-up to the CES plenary session discussion on "Official statistics and climate action".

In October 2021, the Bureau discussed the proposal for follow-up prepared by the Steering Group on Climate Change-Related Statistics. The Bureau approved the proposal, decided to create a Task Force and asked the Steering Group to prepare the terms of reference at its next meeting.

The Bureau reviewed and approved the terms of reference for the Task Force at its February 2022 meeting.

I. BACKGROUND

753. In 2014, the CES Recommendations defined the role of the statistical offices in supporting GHG inventories and producing other climate change-related statistics, and listed many ways in which NSOs could increase their involvement in providing climate-related data. With the adoption of the Paris Agreement and its Work Programme, these recommendations remain valid, but it is now possible to explore new opportunities and identify concrete ways in which NSOs can contribute to achieving national climate change objectives.

754. First of all, from the perspective of the global reporting, data is integral to the implementation of the Paris Agreement. The Agreement relies on pledges (Nationally Determined Contributions) including quantifiable goals measured through indicators (Enhanced Transparency Framework). The collective progress is assessed every five years (the Global Stocktake). However, countries have a lot of flexibility in setting goals, choosing indicators and organizing reporting arrangements. This means there are many opportunities for NSOs to be involved and add value, but their role in any of the Paris Agreement processes is not prescribed.

755. Secondly, from the national perspective, to meet climate objectives, governments must put in place increasingly ambitious mitigation and adaptation policies. The governments need reliable data to design these policies and monitor their success.

756. Finally, the Paris Agreement calls for the governments to enhance climate change education, training, public awareness, public participation and public access to information, recognizing the importance of these steps for enhancing global climate action and achieving global climate objectives. Producing timely and relevant climate change-related statistics is key contribution of NSOs towards addressing this call and increasing public awareness of climate related issues.

757. In October 2021, the Bureau discussed and approved the proposal for a Task Force on the role of NSOs in achieving national climate objectives and requested the Steering Group to prepare the terms of reference for its February 2022 meeting.

II. MANDATE

758. The work of the Task Force on the role of NSOs in achieving national climate objectives is conducted within the framework of CES and its Bureau. The Task Force will be created for a period of two years, after which it will submit a final report. The Steering Group on Climate Change-Related Statistics will guide and oversee the work of the Task Force.

III. OBJECTIVES

759. The main objective of the Task Force is to develop guidance on how NSOs can contribute to achieving national climate objectives. The guidance will identify concrete ways in which NSOs can be involved and showcase what the statistical system already offers to support climate action.

IV. PLANNED ACTIVITIES AND OUTPUTS

760. The main output of the Task Force's work will be a guidance document on how NSOs can contribute to achieving national climate objectives including the following:

(a) Supporting reporting under the Enhanced Transparency Framework of the Paris Agreement and provide input to the Global Stocktake;

(b) Meeting information needs of national climate policymaking, e.g. for planning, measuring and M&E of climate change adaptation, monitoring energy transition, informing just transition, monitoring carbon footprint. The priority areas of national climate policymaking to be covered will be determined by the Task Force;

(c) Informing the broad public about climate-related issues;

(d) Communicating differences between the air emission estimates from SEEA and GHG inventories.

761. The Task Force will carry out the following activities:

(a) Identify relevant resources that are already available (developed by the Steering Group, the Task Force on the Set of Core Climate Change-Related Statistics and Indicators or other groups and organizations);

(b) Organize workshop(s) or consultations with users of data related to achieving climate objectives, such as policymakers, journalists or researchers;

(c) Identify relevant good practices and country examples;

(d) Draft the Guidance document;

(e) Identify areas for further work.

762. The Task Force will focus on the institutional aspect of NSOs' involvement and guidance on the use of existing data, to complement the ongoing methodological work under major global statistical frameworks and initiatives, e.g. the Global Set of Climate Change Statistics and Indicators, SEEA, G20 Data Gaps Initiative or the OECD International Programme for Action of Climate. The Task Force may describe identified data gaps. The Task Force will also aim to showcase how the CES Set of Core Statistics and Indicators could be used in the above areas and take into account the Global Set of Climate Change Statistics and Indicators.

763. In carrying out its activities, the Task Force will cooperate with the UN agencies and other international organizations working in this area. The work should be aligned and coordinated with corresponding efforts under UNFCCC, outcomes of COP26 and with the findings of the Eurostat review of climate change-related statistics.

V. TIMETABLE

764. The Task Force is foreseen to work for two years. The activities of the Task Force are planned for the period from February 2022 to June 2024, according to the following tentative schedule:

Feb-Mar 2022	Establishment of the Task Force and drafting of a work plan with activities, timing and division of work
Mar-Apr 2022	Identifying and reviewing existing resources
May 2022	Organization of workshop(s) or consultation with data users
May-Sep 2022	Identifying good practices and country examples, and developing the outline of the Guidance

Sep 2022	Reporting on the progress to the 2022 Expert Forum for Producers and Users of Climate Change-Related Statistics
Oct 2022 - Aug 2023	Drafting the Guidance
Sep 2023	Presenting the draft Guidance to the 2023 Expert Forum for Producers and Users of Climate Change-Related Statistics
Oct - Nov 2023	Reviewing the Guidance
Dec 2023	Finalizing the Guidance
Jan-Feb 2024	Submission of the Guidance to CES Bureau and discussion at its February 2024 meeting
June 2024	Expected endorsement by the CES plenary session

VI. METHODS OF WORK

765. The Task Force will work primarily via e-mail, collaborative workspace and online conferences. Physical meetings of the Task Force could be organized back-to-back to other meetings attended by a significant number of Task Force members.

VII. MEMBERSHIP

766. The Task Force will include representatives of NSOs, other involved national agencies, academia and international organizations.

767. The following countries, international organizations and research institutes have expressed interest in participating in the Task Force: Armenia, Azerbaijan, Belarus, Canada, Costa Rica, Denmark, Ireland, Italy, Kazakhstan, Netherlands, Poland, Russian Federation, Serbia, Spain, Sweden, Turkey, Ukraine, United Kingdom, ECA, ECLAC, EEA, ESCAP, Eurostat, IEA, IMF, UNEP, UNSD, UNFCCC, Midsummer Analytics and UNECE. The Task Force is open to other countries and organizations that would like to contribute to its work.

768. UNECE will act as Secretariat to the Task Force.

*****

GUIDING PRINCIPLES FOR DRAFTING THE REPORT

Agreed by the Task Force on 2 November 2022

Audience

- The primary audience is NSOs; the Guidance should inspire and support people working in NSOs who want to start or develop their work in this area
- Guidance may also benefit:
    - Data users to inform about what NSOs can offer
    - International organizations as custodians of the relevant frameworks

Value added, scope and content

- Guidance should cover how NSOs can contribute to achieving national climate change objectives through, among others:
    - Producing data
    - Helping standardize data produced by others through harmonizing methods and classifications
    - Building up inventories of existing data
    - Coordinating within the statistical office and with other agencies and ministries
    - Knowledge sharing
    - Improving the accessibility and use of data for informing the public
- Guidance should summarize information that is already in the public domain about the above and translate it to the needs of the primary target audience - the NSOs.
- Guidance should define the main questions about the role of NSOs, identify differences and similarities between countries, and the main challenges.
- Guidance should provide a portfolio of real country examples wherever possible, including:
    - Products implemented by NSOs or in cooperation with other agencies and ministries,
    - Experiences of collaboration, coordination and synergies which brought added value,
    - Case studies,
    - Descriptions of institutional arrangements etc.
- The document should provide guidance specific to climate change-related statistics - avoid generic guidance that could apply to any new topic and make the guidance specific to climate change through examples.
- There might be a different balance between the explanations, practical examples and recommendations in different parts of the Guidance due to differences in topics' maturity, but the structure of chapters should be aligned.
- The Task Force can take up the issue of indicators, e.g., by looking at how Global Set and CES Set indicators can be used and what indicators are currently produced by countries that are useful for reporting under the Paris Agreement or informing climate change mitigation or adaptation policies. But overall, the focus should be more on understanding the themes and the information needs and identifying what indicators are relevant in a particular context rather than going in-depth on particular indicators.
- The Task Force should be guided by the ToR and adapt as needed.
- It is not the purpose to reinvent the wheel, develop new indicators or indicator lists or classifications, or develop a new statistical framework.

Annex 2. Coverage of statistical topics by the frameworks and approaches – summary table

Statistical area	FDES	IPCC	SEEA-CF	Relevance
Demographic and social statistics				
1.1 Population and migration	x	x		closely
1.2 Labour			partly	indirectly
1.3 Education	partly	partly	partly	indirectly
1.4 Health	x	x		closely
1.5 Income and consumption	partly	partly	partly	related
1.6 Social protection			partly	indirectly
1.7 Human settlements and housing	x	x	partly	closely
1.8 Justice and crime				indirectly
1.9 Culture		x	partly	indirectly
1.10 Political and other community activities	partly	x	x	closely
1.11 Time use		partly	partly	related
Economic statistics				
2.1 Macroeconomic statistics			partly	indirectly
2.2 Economic accounts			partly	indirectly
2.3 Business statistics	partly	partly	partly	related
2.4 Sectoral statistics				
2.4.1 Agriculture, forestry, fisheries	partly	partly	x	closely
2.4.2 Energy	partly	partly	x	closely
2.4.3 Mining, manufacturing, construction	partly	partly	x	closely
2.4.4 Transport	partly	partly	partly	closely
2.4.5 Tourism	x		x	closely
2.4.6 Banking, insurance, financial statistics	partly		partly	indirectly
2.5 Government finance, fiscal, public sector	partly	partly	partly	related
2.6 International trade, balance of payments		x	partly	related
2.7 Prices				indirectly
2.8 Labour cost				indirectly
2.9 Science, technology and innovation		x	partly	closely
Environment and multi-domain statistics				
3.1 Environment	x	x	x	closely
3.2 Regional and small area statistics	x	x	x	closely
3.3 Multi-domain statistics and indicators				
3.3.1 Living conditions, poverty, social issues	x	x	partly	closely
3.3.2 Gender and special population groups	partly	x		related
3.3.3 Information society				indirectly
3.3.4 Globalization		x		related
3.3.5 Millennium Development Goals	partly	partly	partly	related
3.3.6 Sustainable development	x	x	x	closely
3.3.7 Entrepreneurship	partly	partly	partly	related

Annex 3. The CES set of core climate change-related indicators

DRIVERS	
1a	Total energy use by the national economy
1b	Total primary energy supply (TPES)
2a	Share of fossil fuels in total energy use by the national economy
2b	Share of fossil fuels in total primary energy supply (TPES)
3	Losses of land covered by (semi-) natural vegetation
4	Total support for fossil fuels in relation to GDP
5a	Total energy intensity of production activities of the national economy
6a	Total CO_2 intensity of energy used in production activities of the national economy
8a	Energy use by resident households per capita
EMISSIONS	
9a	Total greenhouse gas emissions from the national economy
9b	Total greenhouse gas emissions from the national territory
10a	CO_2 emissions from fuel combustion attributable to the national economy
10b	CO_2 emissions from fuel combustion within the national territory
11	Greenhouse gas emissions from land use change (LULUCF)
12	Total greenhouse gas emissions from production activities
13	Greenhouse gas emission intensity of production activities
14	Direct greenhouse gas emissions from households
15	Carbon footprint
IMPACTS	
16	Mean temperature anomaly (compared to climate normal 1961 - 1990)
17	Percentage of land area suffering from unusually wet or dry conditions (Standard Precipitation Index)
18	Level of water stress: freshwater withdrawal as a proportion of available freshwater resources
	Placeholder for an indicator on CC impact on biodiversity
20	Carbon stock in soil
21	Proportion of land that is degraded over total land area
22	Number of deaths and missing persons attributed to hydro-meteorological disasters per 100,000 population
23	Occurrence of extremes of temperatures and precipitation
24	Direct economic loss attributed to hydro-meteorological disasters in relation to GDP
25	Number of people whose destroyed dwellings were attributed to hydro-meteorological disasters
26	Incidence of climate-related vector-borne diseases
27	Excess mortality related to heat
28	Direct agricultural loss attributed to hydro-meteorological disasters
MITIGATION	
29a	Renewable energy share in total energy use by the national economy
29b	Renewable energy share in the total final energy consumption within the national territory
30	Share of climate change mitigation expenditure in relation to GDP
31	Share of energy and transport related taxes in total taxes and social contributions
32	Total climate change-related subsidies and similar transfers in relation to GDP
33	Average trading carbon price
34	Amounts provided and mobilized in United States dollars per year in relation to the continued existing collective mobilization goal of the $100 billion commitment through to 2025
81	Net emissions/removals of carbon dioxide by forest land
ADAPTATION	
35	Share of government adaptation expenditure in relation to GDP
36	Change in water use efficiency over time
39	Proportion of agricultural area under productive and sustainable agriculture
82	Share of green urban areas in the total area of cities
	Placeholder for indicator on CC adaptation by forests

Annex 4. Global Set of Climate Change Statistics and Indicators

Available on the publication's page online.

Annex 5. Summary of interviews with journalists

Available on the publication's page online.

References

Arctic Council (2016). Arctic Resilience Report. M. Carson and G. Peterson (eds). Stockholm Environment Institute and Stockholm Resilience Centre. Stockholm. https://mediamanager.sei.org/documents/Publications/ArcticResilienceReport-2016.pdf.

Atmadja, Stibniati S. and others (2022). How do REDD+ projects contribute to the goals of the Paris Agreement? *Environmental Research Letters*, Vol. 17, No. 4. https://edepot.wur.nl/568301.

Bennett, Nathan J. and others (2019). Just Transformations to Sustainability. *Sustainability*. 11. 3881. 10.3390/su11143881. https://nathanbennett.ca/wp-content/uploads/2019/07/bennett-et-al_2019_just-transformations-to-sustainability.pdf.

BIS, Basel Committee on Banking Supervision (2021). Climate-related risk drivers and their transmission channels. https://www.bis.org/bcbs/publ/d517.pdf.

Brady, Gerry (2022). Use of administrative microdata for climate-related statistics in Ireland. Paper presented at the Expert Forum for Producers and Users of Climate Change-Related Statistics, Geneva, September. https://unece.org/sites/default/files/2022-09/WP5_Session3_Ireland_Administrative_data_for_Climate_Statistics.pdf.

Canadian Centre for Energy Information (2024). Greenhouse gas emissions. 18 January. https://energy-information.canada.ca/en/subjects/greenhouse-gas-emissions.

Central Statistics Office of Ireland and UNECE (2023). Administrative microdata for climate change, energy and environment statistics - webinar note. Paper presented at the Expert Forum for Producers and Users of Climate Change-Related Statistics, Geneva, September. https://unece.org/statistics/documents/2023/08/working-documents/administrative-microdata-climate-change-energy-and.

CMFB (2023). Final report of the Task Force on the statistics on sustainable finance and climate related risks. 13 April. https://circabc.europa.eu/sd/a/7b7397ea-a65c-43b2-97c4-62b79b2592d4/2023-04-13%20-%20Report%20of%20the%20Task%20Force%20on%20statistics%20on%20sustainable%20finance%20and%20climate%20related%20risks.pdf.

Craft, Brianna and Susannah Fisher (2018). Measuring the adaptation goal in the global stocktake of the Paris Agreement. *Climate Policy*. 18(9): pp.1203-1209.

Deutsche Gesellschaft für Internationale Zusammenarbeit (GIZ) and International Institute for Sustainable Development (2018). *Alignment to Advance Climate-Resilient Development.* NAP Global Network. https://napglobalnetwork.org/wp-content/uploads/2018/08/napgn-en-2018-alignment-to-advance-climate-resilient-development-overview-brief.pdf.

Dilling L. and others (2019). Is adaptation success a flawed concept? *Nature Climate Change* (9): pp.570-574. https://doi.org/10.1038/s41558-019-0543-4.

ESCAP, Expert Group on Disaster-related Statistics in Asia and the Pacific (2018). *Disaster-related Statistics Framework*. Bangkok. ESCAP/CST/2018/CRP.2. https://stat-confluence.escap.un.org/display/TWG/DRSF%3A+Disaster-related+Statistics+Framework.

European Commission (2019). Communication from the Commission. *The European Green Deal*. Brussels. COM/2019/640 final. https://eur-lex.europa.eu/legal-content/EN/ALL/?uri=CELEX:52019DC0640.

European Statistical System (ESS) (2018). European Statistics Code of Practice for the National Statistical Authorities and Eurostat. Luxembourg. https://ec.europa.eu/eurostat/documents/4031688/8971242/KS-02-18-142-EN-N.pdf/.

European Statistical System Committee (ESSC) (2021). ESS contribution to the European Green Deal. ESSC 2021/47/03/EN. https://ec.europa.eu/eurostat/documents/1798247/13699600/Action+plan+statistics+for+the+European+Green+Deal.pdf/d3f04cbe-c1f3-edfc-d3f8-903f4a5045ec?t=1637322011610.

Eurostat. Statistics Explained. https://ec.europa.eu/eurostat/statistics-explained/index.php?title=Main_Page.

Facioni, Carolina, Maria Clelia Romano and Anna Villa (2023). Citizens' attitudes and behaviours in environmental matters: a gender-based approach. Paper presented at the UNECE Meeting of the Group of Experts on Gender Statistics, Geneva, May. https://unece.org/statistics/documents/2023/04/working-documents/citizens-attitudes-and-behaviours-environmental.

Ferruzza, Angela (2023). Statistical Register of Places: opportunities for climate change and disaster risk related indicators. Presentation at the Expert Forum for Producers and Users of Climate Change-Related Statistics, Geneva, August. https://unece.org/statistics/documents/2023/08/presentations/statistical-register-places-opportunities-climate-change.

Financial Stability Board (FSB) (2023). FSB Roadmap for Addressing Financial Risks from Climate Change Progress report: 2023 progress report. 13 July. https://www.fsb.org/uploads/P130723.pdf.

IEA (2014). *Energy Efficiency Indicators: Fundamentals on Statistics*. IEA. Paris. https://www.iea.org/reports/energy-efficiency-indicators-fundamentals-on-statistics.

_____ (2021). *The Role of Critical Minerals in Clean Energy Transitions.* IEA: Paris. https://www.iea.org/reports/the-role-of-critical-minerals-in-clean-energy-transitions.

_____ (2022a). Addressing gaps in data for energy-related climate indicators: the IEA's perspective. Presentation at the Expert Forum for Producers and Users of Climate Change-Related Statistics, Geneva, September. https://unece.org/statistics/documents/2022/09/presentations/addressing-gaps-data-energy-related-climate-indicators.

_____ (2022b). *Tracking Public Investment in Energy Technology Research: A Roadmap*. IEA. Paris. https://www.iea.org/reports/tracking-public-investment-in-energy-technology-research-a-roadmap.

_____ (2023). *Demand-side Data and Energy Efficiency Indicators*. IEA. Paris. https://www.iea.org/reports/demand-side-data-and-energy-efficiency-indicators.

ILO (2015). *Guidelines for a just transition towards environmentally sustainable economies and societies for all.* ILO. Switzerland. https://www.ilo.org/publications/guidelines-just-transition-towards-environmentally-sustainable-economies.

Ingole, Vijendra and others (2022). Local mortality impacts due to future air pollution under climate change scenarios. *Science of The Total Environment*, Volume 823, 2022, 153832, ISSN 0048-9697. https://doi.org/10.1016/j.scitotenv.2022.153832. https://www.sciencedirect.com/science/article/pii/S004896972200924X.

IPCC (2006). 2006 IPCC Guidelines for National Greenhouse Gas Inventories. Prepared by the National Greenhouse Gas Inventories Programme, Eggleston H.S., Buendia L., Miwa K., Ngara T. and Tanabe K. (eds). Published: IGES, Japan. https://www.ipcc-nggip.iges.or.jp/public/2006gl/.

_____ (2014). Climate Change 2014: Synthesis Report. Contribution of Working Groups I, II and III to the Fifth Assessment Report of the Intergovernmental Panel on Climate Change [Core Writing Team, R.K. Pachauri and L.A. Meyer (eds.)]. IPCC, Geneva, Switzerland. https://www.ipcc.ch/site/assets/uploads/2018/02/SYR_AR5_FINAL_full.pdf.

_____ (2018): *Global Warming of 1.5°C. An IPCC Special Report on the impacts of global warming of 1.5°C above pre-industrial levels and related global greenhouse gas emission pathways, in the context of strengthening the global response to the threat of climate change, sustainable development, and efforts to eradicate poverty.* Masson-Delmotte, V., P. Zhai, H.-O. Pörtner, D. Roberts, J. Skea, P.R. Shukla, A. Pirani, W. Moufouma-Okia, C. Péan, R. Pidcock, S. Connors, J.B.R. Matthews, Y. Chen, X. Zhou, M.I. Gomis, E. Lonnoy, T. Maycock, M. Tignor, and T. Waterfield (eds.). Cambridge University Press, Cambridge, UK and New York, NY, USA, 616 pp. https://doi.org/ 10.1017/9781009157940. https://www.ipcc.ch/sr15/.

_____ (2019). *2019 Refinement to the 2006 IPCC Guidelines for National Greenhouse Gas Inventories.* Calvo Buendia, E., Tanabe, K., Kranjc, A., Baasansuren, J., Fukuda, M., Ngarize, S., Osako, A., Pyrozhenko, Y., Shermanau, P. and Federici, S. (eds). IPCC, Switzerland. https://www.ipcc nggip.iges.or.jp/public/2019rf/index.html.

_____ (2022a). *Climate Change 2022: Impacts, Adaptation, and Vulnerability. Contribution of Working Group II to the Sixth Assessment Report of the Intergovernmental Panel on Climate Change.* H.-O. Pörtner, D.C. Roberts, M. Tignor, E.S. Poloczanska, K. Mintenbeck, A. Alegría, M. Craig, S. Langsdorf, S. Löschke, V. Möller, A. Okem, B. Rama (eds.). Cambridge University Press. Cambridge University Press, Cambridge, UK and New York, NY, USA, 3056 pp., doi:10.1017/9781009325844. https://www.ipcc.ch/report/ar6/wg2/downloads/report/IPCC_AR6_WGII_FullReport.pdf.

_____ (2022b). *Climate Change 2022: Mitigation of Climate Change. Contribution of Working Group III to the Sixth Assessment Report of the Intergovernmental Panel on Climate Change*. P.R. Shukla, J. Skea, R. Slade, A. Al Khourdajie, R. van Diemen, D. McCollum, M. Pathak, S. Some, P. Vyas, R. Fradera, M. Belkacemi, A. Hasija, G. Lisboa, S. Luz, J. Malley, (eds.). Cambridge University Press, Cambridge, UK and New York, NY, USA. doi: 10.1017/9781009157926.002. https://www.ipcc.ch/report/ar6/wg3/downloads/report/IPCC_AR6_WGIII_FullReport.pdf.

_____ (2023a). Climate Change 2023: Synthesis Report. Contribution of Working Groups I, II and III to the Sixth Assessment Report of the Intergovernmental Panel on Climate Change [Core Writing Team, H. Lee and J. Romero (eds.)]. IPCC, Geneva, Switzerland, 184 pp., doi: 10.59327/IPCC/AR6-9789291691647. https://www.ipcc.ch/report/ar6/syr/downloads/report/IPCC_AR6_SYR_FullVolume.pdf.

_____ (2023b). Urgent climate action can secure a liveable future for all. Press release for AR6 Synthesis Report. 20 March 2023. 2023/06/PR. https://www.ipcc.ch/2023/03/20/press-release-ar6-synthesis-report/

Jensen, Peter Rørmose and Bogomil Iliev (2022). Compilation of a consumption-based greenhouse gas account for Denmark using coupled models. Statistics Denmark. December. https://www.dst.dk/Site/Dst/SingleFiles/GetArchiveFile.aspx?fi=868742149393&fo=0&ext=kvaldel.

Just Transition Center and B Team (2018). *Just Transition: A Business Guide*. https://bteam.org/assets/reports/Just-Transition-A-Business-Guide.pdf.

Leiter, Timo and Patrick Pringle (2018). Pitfalls and potential of measuring climate change adaptation through adaptation metrics. In: L Christiansen, Martinez G, and P Naswa (eds.). *Adaptation metrics: Perspectives on measuring, aggregating and comparing adaptation results*. Copenhagen: UNEP DTU Partnership. pp. 29-48.

McKinsey (2021). *Protecting people from a changing climate: The case for resilience.* November. https://www.mckinsey.com/capabilities/sustainability/our-insights/protecting-people-from-a-changing-climate-the-case-for-resilience.

Millenium Assessment (2005). Appendix D: Glossary. In: *Ecosystems and Human Well-being: Current States and Trends. Findings of the Condition and Trends Working Group* [Hassan, R., R. Scholes and N. Ash(eds.)], Millennium Ecosystem Assessment (MA), Island Press, Washington DC, USA, pp. 893–900. https://www.millenniumassessment.org/en/Condition.html.

Morrin, Paul (2022). Irish approach to data stewardship in the environment domain. Paper presented at the Expert Forum for Producers and Users of Climate Change-Related Statistics, Geneva, September. https://unece.org/statistics/documents/2022/09/working-documents/irish-approach-data-stewardship-environment-domain.

Nachtigall, Daniel and others (2022). The Climate Actions and Policies Measurement Framework: A structured and harmonised climate policy database to monitor countries' mitigation action. *OECD Environment Working Papers*, No. 203. Paris: OECD Publishing. https://doi.org/10.1787/2caa60ce-en.

Network of Central Banks and Supervisors for Greening the Financial System (NFGS) (2022). Final report on bridging data gaps. July. https://www.ngfs.net/sites/default/files/medias/documents/final_report_on_bridging_data_gaps.pdf.

OECD (2015). *Recommendation of the Council on Good Statistical Practice*. https://legalinstruments.oecd.org/en/instruments/OECD-LEGAL-0417.

PARIS21 (2024). Mobilising Climate Change Data Ecosystems Framework: A Framework and Toolkit. https://www.paris21.org/sites/default/files/media/document/2024-05/mobilising-climate-change-data-ecosystems.pdf.

Pizarro, Rodrigo and others (2024). GHG Emission Trends and Targets (GETT): Harmonised quantification methodology and indicators. OECD Environment Working Papers, No. 230. Paris: OECD Publishing. https://doi.org/10.1787/decef216-en.

Statistics Netherlands (2010). SEEA as a framework for assessing policy responses to climate change. Paper presented at the Fifth Meeting of the UN Committee of Experts on Environmental-Economic Accounting. New York. June 2010. https://unstats.un.org/unsd/envaccounting/ceea/meetings/UNCEEA-5-13.pdf.

Statistics New Zealand (2020). Approaches to measuring New Zealand's greenhouse gas emissions. 27 August. https://www.stats.govt.nz/methods/approaches-to-measuring-new-zealands-greenhouse-gas-emissions.

Thunus, Olivier (2023). Use of Air emissions indicators – Luxembourg experience. Presentation at the Regional training on Air Quality and Emissions to air statistics and indicators, Geneva, May. https://unece.org/environment/documents/2023/05/presentations/use-air-emissions-indicators-luxembourg-experience.

Tompkins, Emma L. and others (2018). Documenting the state of adaptation for the global stocktake of the Paris Agreement. WIREs Climate Change. 9:e545. https://doi.org/10.1002/wcc.545.

UN Women, Women Count, and United Nations (2022). Progress on the Sustainable Development Goals: The Gender Snapshot 2022. New York: UN Women and United Nations. https://www.unwomen.org/en/digital-library/publications/2022/09/progress-on-the-sustainable-development-goals-the-gender-snapshot-2022

UNECE (2014). *Conference of European Statisticians Recommendations on Climate Change-Related Statistics*. Geneva. https://unece.org/statistics/publications/recommendations-climate-change-related-statistics.

_____ (2017). Example Road Maps for Developing Climate Change-Related Statistics. https://unece.org/fileadmin/DAM/stats/documents/ece/ces/2017/Road_maps_for_climate_change_statistics.pdf

_____ (2019a). Generic Statistical Business Process Model. https://statswiki.unece.org/display/GSBPM

_____ (2019b). *Guidance on Modernizing Statistical Legislation*. United Nations publication. https://unece.org/statistics/publications/guidance-modernizing-statistical-legislation.

_____ (2019c). *Recommendations on the Role of Official Statistics in Measuring Hazardous Events and Disasters*. United Nations publication. https://unece.org/statistics/publications/recommendations-role-official-statistics-measuring-hazardous-events-and-disasters.

_____ (2020). In-depth review of the role of the statistical community in climate action. https://unece.org/statistics/ces/climate-action.

_____ (2021a). Conference of European Statisticians' Set of Core Climate Change-related Indicators and Statistics Using the System of Environmental-Economic Accounting. https://unece.org/statistics/publications/CES-set-of-core-climate-change-related-indicators.

_____ (2021b). Implementation Guidelines for the Conference of European Statisticians' Set of Core Climate Change-related Indicators and Statistics using the System of Environmental- Economic Accounting. https://unece.org/statistics/publications/implementation-guidelines-CES-core-set-CCRSI.

_____ (2022). *The Fundamental Principles of Official Statistics: Marking the Anniversary*. https://w3.unece.org/fpos30/ and https://unece.org/fpos30.

_____ (2023a). *Data stewardship and the role of national statistical offices in the new data ecosystem*. United Nations publication. https://unece.org/statistics/publications/data-stewardship-and-role-national-statistical-offices-new-data-ecosystem.

_____ (2023b). *Measuring hazardous events and disasters: set of core disaster-risk-related indicators* https://unece.org/statistics/documents/2023/05/working-documents/measuring-hazardous-events-and-disasters-set-core-0.

UNECE CES (2020). Conclusions of the Chair of the Expert Forum for Producers and Users of Climate Change-Related Statistics. Geneva. https://unece.org/fileadmin/DAM/stats/documents/ece/ces/ge.33/2020/mtg3/2020_EF_Conclusions.pdf.

_____ (2021). Report of the 2021 Expert Forum for Producers and Users of Climate Change-Related Statistics. Geneva. https://unece.org/statistics/documents/2022/08/reports/report.

_____ (2022). Conclusions of the 2022 Expert Forum for Producers and Users of Climate Change-Related Statistics. Geneva. https://unece.org/statistics/documents/2022/12/reports/conclusions-2022-expert-forum.

UNEP (2016). Definitions and Concepts: Background Note. Inquiry Working Paper 16/13. https://www.unep.org/resources/report/definitions-and-concepts-background-note-inquiry-working-paper-1613.

_____ (2017). *The Adaptation Gap Report 2017*. Nairobi. https://www.unenvironment.org/resources/adaptation-gap-report-2017.

_____ (2019). Measuring Fossil Fuel Subsidies in the Context of the Sustainable Development Goals. Nairobi: UN Environment. https://wedocs.unep.org/bitstream/handle/20.500.11822/28111/FossilFuel.pdf?sequence=1&isAllowed=y.

_____ (2022). *Emissions Gap Report 2022: The Closing Window – Climate crisis calls for rapid transformation of societies*. Nairobi. https://www.unep.org/emissions-gap-report-2022.

UNEP and IUCN (2018). *Gender and environment statistics: Unlocking information for action and measuring the SDGs*. Nairobi: UN Environment. https://www.unep.org/resources/report/gender-and-environment-statistics-unlocking-information-action-and-measuring-sdgs.

UNESCO and UNFCCC (2016). *Action for Climate Empowerment: Guidelines for accelerating solutions through education, training and public awareness.* Paris: UNESCO. https://unfccc.int/files/cooperation_and_support/education_and_outreach/application/pdf/action_for_climate_empowerment_guidelines.pdf.

UNFCCC (1992). United Nations Framework Convention on Climate Change. https://unfccc.int/resource/docs/convkp/conveng.pdf.

_____ (1997). Kyoto Protocol to the United Nations Framework Convention on Climate Change. https://unfccc.int/resource/docs/convkp/kpeng.pdf.

_____ (2015). Paris Agreement. https://unfccc.int/sites/default/files/english_paris_agreement.pdf.

_____ (2016a). Aggregate effect of the intended nationally determined contributions: an update. Synthesis report by the secretariat. 2 May. FCCC/CP/2016/2. https://unfccc.int/resource/docs/2016/cop22/eng/02.pdf.

_____ (2016b). Just transition of the workforce, and the creation of decent work and quality jobs. Technical paper. UNFCCC. https://unfccc.int/sites/default/files/resource/Just%20transition.pdf.

_____ (2018). Report of the Conference of the Parties serving as the meeting of the Parties to the Paris Agreement on the third part of its first session, held in Katowice from 2 to 15 December 2018. FCCC/PA/CMA/2018/3/Add.1-2. https://unfccc.int/documents/193407 and https://unfccc.int/documents/193408.

_____ (2021a). Climate Action Pathway: Climate Resilience - Executive Summary. https://unfccc.int/sites/default/files/resource/ExecSumm_Resilience_0.pdf.

_____ (2021b). Report of the Conference of the Parties serving as the meeting of the Parties to the Paris Agreement on its third session, held in Glasgow from 31 October to 13 November 2021. Addendum Part two: Action taken by the Conference of the Parties serving as the meeting of the Parties to the Paris Agreement at its third session. FCCC/PA/CMA/2021/10/Add.2 https://unfccc.int/documents/460951

_____ (2022a). Nationally determined contributions under the Paris Agreement. Synthesis report by the secretariat. 26 October. FCCC/PA/CMA/2022/4 https://unfccc.int/documents/619180.

_____ (2022b). Reference Manual for the Enhanced Transparency Framework under the Paris Agreement Understanding the enhanced transparency framework and its linkages. Available at: https://unfccc.int/sites/default/files/resource/v2_ETFreferencemanual.pdf.

_____ (2022c). Synthesis report for the technical assessment component of the first global stocktake: State of adaptation efforts, experiences and priorities. Prepared by the secretariat under the guidance of the co-facilitators of the technical dialogue of the first global stocktake. 29 April. https://unfccc.int/sites/default/files/resource/Synthesis%20report%20on%20the%20state%20of%20adaptation%20efforts%2C%20experiences%20and%20priorities.pdf.

_____ (2023). General and cross-cutting aspects for the technical expert review under the enhanced transparency framework under the Paris Agreement. https://unfccc.int/sites/default/files/resource/General_and_cross_cutting_aspects_for_the_TER_under_ETF_PA.pdf.

UNFCCC Adaptation Committee (2019). 25 Years of Adaptation under the UNFCCC. Report by the Adaptation Committee. https://unfccc.int/sites/default/files/resource/AC_25%20Years%20of%20Adaptation%20Under%20the%20UNFCCC_2019.pdf.

_____ (2020). Data for adaptation at different spatial and temporal scales. Technical paper by the Adaptation Committee. https://unfccc.int/sites/default/files/resource/AC%20adaptation%20data%20full.pdf.

_____ (2022). Synthesis report for the technical assessment component of the first global stocktake. 12 May. https://unfccc.int/documents/476683.

UNFCCC Subsidiary Body for Implementation (2012). Article 6 of the Convention. FCCC/SBI/2012/L.47. https://unfccc.int/resource/docs/2012/sbi/eng/l47.pdf.

UNFCCC Least Developed Countries Expert Group (2012). *National Adaptation Plans. Technical guidelines for the national adaptation plan process*. December. Available at https://napcentral.org/nap-guidelines

_____ (2018). Considerations regarding vulnerable groups, communities and ecosystems in the context of the national adaptation plans. December. https://unfccc.int/sites/default/files/resource/Considerations%20regarding%20vulnerable.pdf.

UNFCCC Standing Committee on Finance (2022). Fifth Biennial Assessment and Overview of Climate Finance Flows. https://unfccc.int/sites/default/files/resource/J0156_UNFCCC%20BA5_2022_Report_v4%5B52%5D.pdf

_____ (2023). Report on clustering types of climate finance definitions in use. https://unfccc.int/sites/default/files/resource/Definitions_final_231117%20BLS23393%20UCC%20Climate%20Finance.pdf.

UNISDR (2015). Sendai Framework for Disaster Risk Reduction 2015–2030. https://www.undrr.org/quick/11409.

United Nations (2014). Fundamental Principles of Official Statistics. A/RES/68/261. https://unstats.un.org/unsd/dnss/gp/fp-new-e.pdf.

_____ (2015). *Transforming our world: The 2030 Agenda for Sustainable Development.* A/Res/70/1. https://sustainabledevelopment.un.org/content/documents/21252030%20Agenda%20for%20Sustainable%20Development%20web.pdf.

_____ (2016). *Integrating a Gender Perspective into Statistics.* https://unstats.un.org/unsd/demographic-social/Standards-and-Methods/files/Handbooks/gender/Integrating-a-Gender-Perspective-into-Statistics-E.pdf.

_____ (2017). *Framework for the Development of Environment Statistics* (FDES 2013). https://unstats.un.org/unsd/envstats/fdes.cshtml.

_____ (2018). *International Recommendations for Energy Statistics.* https://unstats.un.org/unsd/energystats/methodology/documents/IRES-web.pdf.

_____ (2019). The Sustainable Development Goals Report 2019. https://unstats.un.org/sdgs/report/2019/The-Sustainable-Development-Goals-Report-2019.pdf.

_____ (2024). *Global Set of Climate Change Statistics and Indicators: Implementation Guidelines.* https://unstats.un.org/unsd/envstats/climate_change/Implementation_Guidelines.pdf.

United Nations and others (2014). *System of Environmental-Economic Accounting 2012: Central Framework.* United Nations. https://seea.un.org/content/seea-central-framework.

_____ (2021). *System of Environmental-Economic Accounting – Ecosystem Accounting*. White cover publication, pre-edited text subject to official editing. https://seea.un.org/ecosystem-accounting.

United Nations General Assembly (2016). Report of the open-ended intergovernmental expert working group on indicators and terminology relating to disaster risk reduction. A/71/644. https://www.undrr.org/publication/report-open-ended-intergovernmental-expert-working-group-indicators-and-terminology.

_____ (2017). Work of the Statistical Commission pertaining to the 2030 Agenda for Sustainable Development. A/RES/70/313. https://undocs.org/A/RES/71/313.

UNSD (2022). Background document to the report of the Secretary-General on climate change statistics (E/CN.3/2022/17), Global Set and metadata. Available at https://unstats.un.org/unsd/statcom/53rd-session/documents/BG-3m-Globalsetandmetadata-E.pdf.

_____ (2023). Climate Change Statistics and Indicators Self-Assessment Tool (CISAT). Available at https://unstats.un.org/unsd/envstats/climate%20change/cisat.cshtml.

_____ (2024). Tier Classification for Global SDG Indicators as of 6 March 2024. https://unstats.un.org/sdgs/iaeg-sdgs/tier-classification/.

Werkhoven, Thom (2022). European Green Deal in the Netherlands. Paper presented at the Expert Forum for Producers and Users of Climate Change-Related Statistics, Geneva, September. https://unece.org/statistics/documents/2022/09/working-documents/european-green-deal-netherlands.

WHO (2023). Fact sheets: Climate Change. 12 October. https://www.who.int/news-room/fact-sheets/detail/climate-change-and-health.

World Wildlife Fund for Nature (2021). Just transitions towards a nature-positive economy: Learning from stories of change. March. https://wwfint.awsassets.panda.org/downloads/just_transitions_towards_a_nature_positive_economy_summary_1.pdf.